'*Performing Climates* speaks brilliantly to the new zeitgeist of artistic and cultural thinking that is showing ways of living with climate change empathetically and with attention to multiple cultural knowledge systems and artistic sensibilities. Paterson and Stevens show how to live with sensitive attention to the everyday, situated inextricably between thinking, creativity and activism.'

Peter Eckersall, *City University of New York*

Performing Climates

Performing Climates features 14 interconnected chapters exploring theatre and performance's relationship with more-than-human elements at a time of climate emergency. The book argues that Western performance – how we conceive of it, as well as how we train and educate people in and about it – needs to reorient its ways of making and thinking about itself to reconsider patterns of breakdown, decay and renewal happening on and off stage in a literal play of cells and particles. *Performing Climates* examines live performance as a *uniquely compostable artform,* formed by sonic vibrations and movements of air and matter, more-than-human elements, composition and decomposition. This book will appeal to undergraduate audiences, postgraduate scholars and performance studies colleagues, offering exciting possibilities for reconsidering theatre and performing in an age of crisis.

Eddie Paterson is Associate Professor in Creative Writing at the University of Melbourne.

Lara Stevens is Lecturer in English at Charles Sturt University, Australia.

Routledge Advances in Theatre & Performance Studies

This series is our home for cutting-edge, upper-level scholarly studies and edited collections. Considering theatre and performance alongside topics such as religion, politics, gender, race, ecology, and the avant-garde, titles are characterise by dynamic interventions into established subjects and innovative studies on emerging topics.

Researching Popular Entertainment
Edited by Kim Baston & Jason Price

The Dancer's Handbook
A Practical Guide to Reimagining Dance, Body, and Society by OFEN Co-Arts
Gala Moody and Michael Carter

Staging Muslims in Britain
Playwrighting, Performance, and Representation
Önder Çakırtaş

Performing Violence
Limits and Transformative Means in Staged Violence
Davide Giovanzana

Performing Corporate Bodies
Multinational Theatre in Global India
Sarah Saddler

Performing Climates
Eddie Paterson and Lara Stevens

For more information about this series, please visit: www.routledge.com/Routledge-Advances-in-Theatre–Performance-Studies/book-series/RATPS

Performing Climates

Eddie Paterson and Lara Stevens

With a Contribution from Angie Abdilla

Routledge
Taylor & Francis Group
LONDON AND NEW YORK

First published 2025
by Routledge
4 Park Square, Milton Park, Abingdon, Oxon OX14 4RN

and by Routledge
605 Third Avenue, New York, NY 10158

Routledge is an imprint of the Taylor & Francis Group, an informa business

British Library Cataloguing-in-Publication Data
A catalogue record for this book is available from the British Library

ISBN: 978-1-032-22556-2 (hbk)
ISBN: 978-1-032-22560-9 (pbk)
ISBN: 978-1-003-27308-0 (ebk)

DOI: 10.4324/9781003273080

Typeset in Sabon
by KnowledgeWorks Global Ltd.

Contents

Figures

Author Biographies

Angie Abdilla, *palawa*, is Professor of Cybernetics at the Australian National University, Founder, Director of Old Ways, New and co-founder of the Indigenous Protocols and Artificial Intelligence working group (IP//AI).

Eddie Paterson is Associate Professor in Creative Writing at the University of Melbourne. He is the author of *The Contemporary American Monologue* and co-author *of Once Upon a Pixel: Storytelling and Worldbuilding in Videogames*, with Tim Williams and Will Cordner.

Lara Stevens is Lecturer in English at Charles Sturt University, Australia. She is the author of *Anti-war Theatre After Brecht* and co-editor *of Feminist Ecologies: Changing Environments in the Anthropocene*, with Peta Tait and Denise Varney.

Authors' Note

This book was written on the unceded lands of the Dhudhuroa, Waywurru, Gadigal and Wurundjeri Woi Wurrung peoples. We acknowledge the Traditional Owners and pay respects to their Elders, past, present and emerging.

Acknowledgements

Thanks to Melinda Hetzel for early research assistance and Laura Hussey and the team at Routledge for guiding the production of this book. We are indebted to Dr Kyle Harvey for his meticulous editorial contribution. This book was developed with the support of Denise Varney, Peter Eckersall and Peta Tait on the research project 'Towards an Australian Ecological Theatre'. Thanks to the School of Culture and Communication and the Arts Faculty at the University of Melbourne for publication support. Previous versions of some chapters were published elsewhere; we gratefully acknowledge the publishers, editors and peer-reviewers of those volumes. This research was partly funded by the Australian Government through the Australian Research Council (ARC). The views expressed herein are those of the authors and are not necessarily those of the Australian Government or the ARC.

1 Performing Climates

> We are compost, not posthuman; we inhabit the humusities, not the humanities ... Beings—human and not—become with each other, compose and decompose each other, in every scale and register of time and stuff in sympoietic tangling, in earthly worlding and unworlding.
>
> (Haraway in Tsing et al. 2017: M45)

Living with Mess

Performance is and has always been a mess, a bloody mess as we say in Australia (and in Forced Entertainment). And that's what draws us to it. Performance is a mess of matter, climates, things, actors and affects: neither a dramatic nor postdramatic theatre but a network of dramaturgical elements; a site of birth and death, decay and renewal. Performance is a place of *composting* and mulching (and Eddie loves mulching). This book of 14 interconnected chapters, each oriented around a more-than-human element or being (ice, ants, ears, mermaids, etc.), will resituate Western theatre in relation to the biological, cellular and bacterial conditions which govern our planetary existence.

We think that Western performance – how we conceive of it as well as how we train and educate people in and about it – needs to reorient its ways of making and thinking about itself to reconsider patterns of breakdown, decay and renewal happening on- and off-stage in a literal play of cells and particles. These chapters are our way of tracing the impact of climates on our work and life in the theatre, in the garden, in our family and in the world. *Performing Climates* is a book about messy relations: relations between people of different places, histories and backgrounds; relations between human and more-than-human entities; relations between built environments and natural ones; and relations between the past, the present climate emergency and the future of the planet.

Performance is always rooted in a place with its own microclimate connected to broader systems and spheres of relation. But what does it mean to make, watch and think about performance in a climate emergency? We

DOI: 10.4324/9781003273080-1

asked each other this question as we stood in a rehearsal room that was filling with bushfire smoke. Do we have the time or capacity at this moment to stand around in air-conditioned black boxes staring in wonder at the dancing dust particles in the air? Should we have racked up all those carbon miles to fly to Japan to watch a performance of robots when the country was still reeling from the fallout of the Fukushima nuclear disaster? Are we crying Lear as the world burns? As we relocated from the Australian city of Naarm (Melbourne) to the countryside, how was the changing environment altering our perception of the natural world we were (re)discovering? How did the growing pile of toys and other objects that seem to accompany having children further deepen our position as privileged settlers living on stolen Indigenous land? How might performance reengage with the messiness of the world off-stage?

Live performance is a form of art that explicitly engages with the climates around it, but these climates are not always well understood or even visible. While theatre has always been composed of human and more-than-human entanglements, *Performing Climates* explores interactions in performance from the micro (elemental particles) to the macro (climate change). In these rapidly changing environmental and social conditions, we need to reinterrogate the role of theatre and performance in the fabric of our global cultures. This book argues that live performance, being composed of living and dead skin cells, sonic vibrations and movements of air and matter, composition and decomposition, is a *uniquely compostable art form*. Considering performing climates in an age of crisis requires us to reenvisage performance in and of decay, in and of atmospheres, and in and outside of human scale and human perception (Figure 1.1).

This book makes no apology for being a sprawling, messy book. 'What is mess' writes Isabelle Stengers, but 'the irreducible and always embedded interplay of processes, practices, experience and ways of knowing and valuing that makes up our common world' (2018: 120). We lead sprawling, messy lives. ('We have children!', cry the parents. 'We are still children!', cry the parents to their parents. Aren't families always the messiest of relations?) *Performing Climates* mixes autobiographical stories, and the places they play out, merging them into analysis of artworks that view the natural and the synthetic as performing subjects alongside a series of images. We thank our families, particularly our two children, as well as colleagues, students and friends near and far for being part of this process. We also attempt to situate ourselves and our biases in the places we saw particular performances or events.

In this book, we try to live by our methodology. In so doing, we draw on and owe a debt to the anti-colonial methodologies of scholars such as Max Liboiron (2021), Aileen Moreton-Robinson (2000), the Gay'wu Group of Women (2019), Mary Graham (1999), Irene Watson (2014), Dwayne Donald (2010) and Kim TallBear (2014), all of whom remind us that our research does not occur in a vacuum, but rather in relations to the places

Figure 1.1 *Bloody Mess* Empty Stage, Forced Entertainment © 2020.
Photo: Hugo Glendenning.

we live and their histories, human and more-than-human, elemental and geological. We are indebted to First Nations colleagues Angie Abdilla – who is co-author of Chapter 14, *Tree* – Tony Birch, Brian Martin, Latai Taumoepeau and Alexis Wright for so generously sharing their time, thoughts and expertise during the writing of this book. We also acknowledge the deep influences of Freya Mathews (1991), Val Plumwood (1993), Deborah Bird Rose (1992) and Ariel Salleh (1997). This book would not be the same without the Plumwood Committee, particularly Affrica Taylor and Plumwood Mountain itself, for the embodied lessons they have so generously offered up. We are grateful for our relationships with artists Lucy Bleach, Hanna Cormick,[1] Maria Fernanda Cardoso, Karen Finley, Gregory Lorenzutti, Norie Neumark, Kris Verdonck and Robert Walton who have given up their time and energy to entrust us with their stories and with our serious and playful explorations of their work.

Canadian Métis scholar Max Liboiron teaches us that respectful and reciprocal relations in all modes of life, including in scholarship, begin with a recognition of the smallest particularities of the places from which we work. These places inform and infuse who we are and what we do, not necessarily in a spiritual sense, though it might be that too, but materially. Liboiron writes: 'methodology is a way of being in the world and ways of being are tied up in obligation' (2021: 1). Our obligation is to our teachers who show us the blind spots in our own thinking and ask us to tread carefully as we reconsider relations with human, more-than-human, living and nonliving elements and our place in the world.

As a consequence, we are not interested in performing what Donna Haraway calls the 'god trick' to assert our scholarly authority and hide our emotional, spiritual and material connection to our subject matter (1988: 581). Instead, we recognise our particular relations to performance and people. We try to acknowledge all we have learned, and continue to learn, from our parents and family, our friends and local community and Indigenous, non-Western and feminist philosophies that teach us that place-based relationality is what fosters responsibility to protect and nurture Country. Methodology matters, even and especially if it is not flawless, symmetrical or easy for us or for you, dear readers.

Ours, Not Ours

We live on unceded Indigenous land. We live on the ancestral lands of the Dhudhuroa and Wayrurru peoples. We live in the north-east of a southern state in Australia, in a place called Wooragee. Wooragee is said to be derived from the Dhudhuroa term for 'peppermint tree' or Wurritba (Blake and Reid 2002). Most likely Wooragee refers to the *Eucalyptus dives*, commonly known as the Broad-leaved Peppermint or Blue Peppermint Gum, a peppermint-scented species endemic to the stony soils of higher places in South Eastern Australia. Directly outside of our kitchen window grows a peppermint tree. However, this tree is not a peppermint gum, but a peppermint willow, *Agonis flexuosa*. The *Agonis* is a species native to a place that is around 3000 kilometres away in Western Australia. To arrive here, it has not travelled on the wind. Yet, the appearance of this tree doesn't seem to be a random coincidence. We can only imagine that it has been grown from seed, tended and raised, transplanted and transported across the country. Perhaps it was purchased at a plant nursery nearby, and because its common English name is the peppermint tree, perhaps it was even planted here in a well-meaning but misguided gesture of recognition of the Dhudhuroa people, their language and connection to this place. Like us, the misplaced Wooragee *Agonis* lives on land that is not its own.

Let us set the scene in a different way. We are a couple with two young children. We are two middle-aged scholars, theatre nerds, one obsessive gardener, both animal lovers, settler Australians, ecofeminist philosophers and activists. We have relocated our lives to an alpine region where Traditional Owners have sustained what Australian Indigenous communities refer to as 'Country' for over 65,000 years. Indigenous knowledge teaches us that Country and the reciprocal relations between First Peoples and their places are bound to what settlers might see as cultural and artistic practices, but which are so deeply integrated that they are inextricable for Indigenous cultures. As Ann McGrath and Laura Rademaker write, in the introduction to *Everywhen: Australia and the Language of Deep History*:

> The land is the theater of storytelling and itself the storyteller – it provides proof, containing relics of the deep past, and it activates a living history. Words spoken or sung and dances reenacted in that landscape really matter, for there is an iterative relationship among the place, the present generations, and ancestral being. Country enhances the Indigenous sense of continuity. The land remains a creative, animating force, generating people, animals and geographies.
>
> (2023: 15–16)

As settler Australians, we are awkwardly positioned to write about that which is both ours and not ours. Walking with and talking to Traditional Owners leads us to acknowledge that we are bound up with, nourished and indebted to Dhudhuroa Wayrurru Country and its custodians, and that we are now part of the custodianship of this place, with deep and lasting responsibility and obligations (Birch 2019b).

What ancient knowledge is available to take and use for our analysis of performance is always fraught territory, and we take guidance where it is offered, act in good faith and know that at times it will be unavoidably messy. We recognise that we walk a fine line and so defer to the instructions that come with the shared knowledge in each particular instance. We are reminded that colonisation is a process of erasing language, culture and species. We arrive with the attitude that to attempt to recuperate and engage earnestly, playfully and deeply with Indigenous knowledge and to make mistakes, acknowledge and learn from Traditional Owners is better than continuing the colonial project of ignoring or eliding such knowledge because we fear being extractive or disrespectful.

In her recent novel *Praiseworthy*, Indigenous Australian author and Waanyi woman Alexis Wright writes playfully of the vital importance of listening to Indigenous people, 'the world's greatest human survivors':

> Extinction-less tempest people from enduring one million storms, come from the ark of infinity right down to the last baby. They were now forced to speak, and yelled that they knew a thing or two about being trodden upon, and of being more on trend than the rest of humanity about how to look after the future.
>
> (2023: 2)

In taking up the knowledge offered by First Nations people, we acknowledge our complicity and responsibility in protecting and fairly representing that knowledge to the best of our ability (Neale 2022: 1).

Performing Climates mixes the autobiographical and analytical alongside a series of images and lyrical imaginings. Anna Lowenhaupt Tsing, author of *The Mushroom at the End of the World*, uses similar mixed modes as a means of capturing the 'lively activities of all beings, human and not human' that have historically been left to fabulists and non-Western storytellers (2015: v). We ask questions, tell stories and offer moments of poetic illustration and nontextual material in the hope of reckoning with being in the world with animals, rivers, rocks, plants, atmospheres, memory and spirits. Wright describes the inspection of the golden beetle (*Anoplognathus aureus*) as it moves across Country:

> It seems to be imagining the world through the fine old music of the trees sing-talking back and forth while relaying all the storylines, the sounds of infinity reaching through a multitude of tracks over vast distances. It crawls through the leaf litter on the ground, along twigs, turns over soil, moves pebbles around while searching for the tiny devil mob. It finds all of those spirits with antennae that do not stop moving.
>
> (2023: 14)

Remember the image of the misplaced peppermint tree. This image reminds us that when we sat down to write this book, we were new and returned inhabitants to the ancestral homelands of the Dhudhuroa people. Dhudhuroa Country was stolen by colonial forces in the eighteenth century. By the nineteenth century, its invading inhabitants became very wealthy from gold prospecting, a pattern of theft and destructive practices of extraction and agricultural mismanagement that has defined Australian colonialism and continues to this day.

Liboiron writes that settler scholarly practices can be extractive too, reading Indigenous knowledge with a quarrying mindset, mining it for what can be 'used' for settler-colonial purposes and self-advancement (2021: 35–37). As settlers on these lands working with Indigenous philosophy, we are accountable for building reconciliation and acknowledging the lineage of First Nations thought, as a mode of rectifying exploitative practices and regenerating place and people. Our friend and colleague, the Indigenous Australian author, scholar and activist Tony Birch confirms that 'through an intellectual, cultural and, for some, a spiritual attachment to place and country', Indigenous and non-Indigenous people can promote collective responsibility for future ecological care (2016: 381).[2]

In this way, our work and life exist in what Liboiron calls a 'place-based relationality' (2021) after the work of Mary Graham (2009), Aileen Moreton-Robinson (2015), Shawn Wilson (2008), Alex Wilson (2008) and Dwayne Donald (2012). The *Agonis* appears several times in this book. Each time it offers a new intermingling image: a tree breathing in a biodome, the symbiotic relation between fungi, bacteria and root, and a shimmering example of photosynthesis (Rose 2017), but the peppermint tree always brings us back to a place that is ours and not ours.

In exploring the meaning and shapes of performing climates, we seek solidarity with the more-than-human, memories of the past and hopes for the future. Eddie grew up here. He is returning to an idea of 'home' after 22 years away. This place, Wooragee, is deeply knitted into his being and his dreams. It is a place of human and more-than-human ghosts. Eddie passed his childhood playing in this dry and spectacular bush, swimming in its creeks and planting trees along their banks – learning from Elders like the late-Bangerang (Pangerang) artist and educator, Eddie 'Kookaburra' Kneebone who was generous with his knowledge of Country. While Eddie holds no rightful claim over these lands, they hold a rightful claim over him.

Lara is also marked by the past and present. As the child of Sri Lankan and British immigrants, Lara's connections to place extend from a suburb of Melbourne on the banks of the Birrarung (Yarra) river and the lands of the Wurundjeri Woi-Wurrung peoples to post World War II South London, and the mercantile networks of the Sri Lankan Burghers, descended from the fifteenth- to eighteenth-century Portuguese and Dutch setters of what Europeans then called Ceylon. As a child of migrants, she continues in that recent familial tradition of moving and setting down roots in places where the newcomer might be considered an outsider but who arrives with fresh eyes, open to new learning, with senses not yet blunted by habit or preconceived ideas.

Compost as Performance

The basis of this book's conceptual framework is grounded in compost (pun intended and achieved) as a cycle of decomposition that generates something vital, rich and useful. We are interested in analysing performance not as an insulated, exclusive art form predicated on human centrality but as a process that celebrates the transformative properties of waste – bacteria, debris and breakdown – composting and mulching within a larger network of bacteria, fungi and microbes embedded in the skin, air, soil and interacting with cellular networks and atmospheric conditions (Figure 1.2).

First Nations' knowledge and cosmologies pre-empt the Western philosophical notion of human and more-than-human co-becoming by many thousands of years. For many Indigenous artists, performance-makers and philosophers, the ideas animating this book will be old news! Concepts of relationality, the practice of making performances that centre around something other than the human, and learnings drawn from deep knowledge of a particular place and climate, is both ancient wisdom and everyday knowledge for many Aboriginal people.[3] In his essay 'Walking and Being', Tony Birch conveys the foundational 'connection between humans and earth' through a Kulin creation story. In the story, the first men and women are sculpted from the clay and waters of riverbanks and creeks – humans are literally and spiritually composed of Country (2019a).

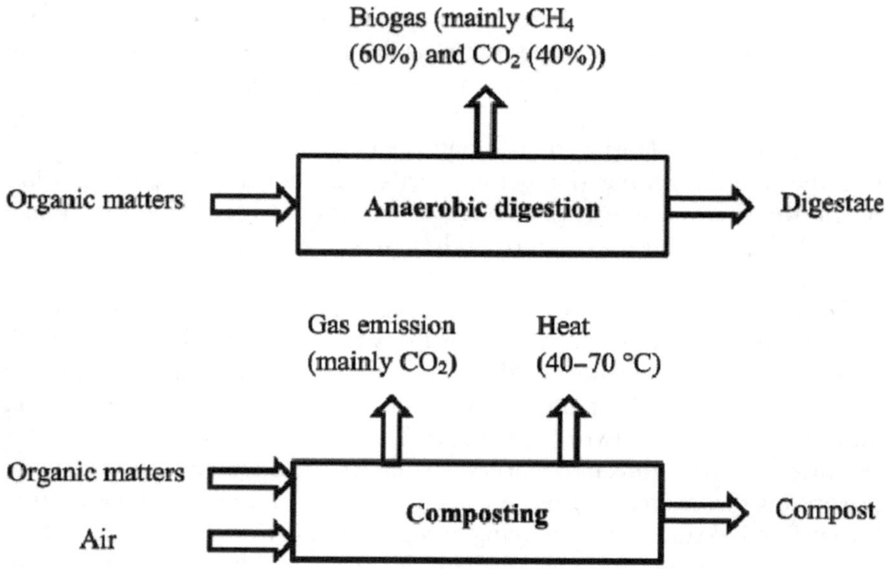

Figure 1.2 Composting.

Indeed, it is through ancient practices of listening to and caring for Country that Australian Indigenous cultures have retained the idea of an animate and lively world of interconnected systems. The Gay'wu Women write:

> People might think that the land doesn't speak, nor the rivers, the animals or the winds, but they do. We feel the land, the sea, and the rivers, and they tell us many things.
>
> (2019: xxviii)

Within knowledge systems where the lands, winds and oceans speak, the notion of performing climates requires no imaginative leap. The things that are told through the land are layered through family histories, kinships, spiritual relations, defining places that hold particular meanings or where particular practices were or should be carried out, many of which are not accessible to settlers.

In this book, we draw upon the Australian Aboriginal term Country to keep us grounded in the place we live, even when our performance examples are global.[4] As Waanyi author Wright reminds us, '*We studied everything you needed to know about surviving from the biggest library in the world – country*' (2023: 2, italics in original). We also draw on recent Western philosophical accounts of the integrated ecologies of human and more-than-human life, including concepts such as 'vibrant matter' (Bennett 2010), 'intra-active' relations (Barad 2007), 'transcorporeality' (Alaimo 2014), the 'planthropocene' (Myers 2017) and 'situated knowledges' (Haraway 1988). Such accounts demonstrate that art and performance has the potential to sustain Jane Bennett's 'thing-power' (2010) and to visualise networks such as what Bruno Latour calls a 'parliament of things' in which groupings of nonhumans, quasi-objects and hybrids are granted visibility and rights (1993: 145).

While we are building our three-bay composting structures, we recall Jennifer Hamilton and Astrida Neimanis arguing that composting is not only a material labour for enhancing the nutrients in the soil but also a scholarly practice. Their work attempts a reclamation of the under/unvalued or ignored foundations to environmental philosophy in feminist scholarship or the writings of other marginalised groups such as Women of Colour or Indigenous Women (2018). Following their example, we draw on and acknowledge, through citation, the undervalued work done in this field, particularly the influence of First Nations and feminist thought on theoretical and artistic modes of performing climates.

Our concept of performing climates is far less complex and culturally significant than Indigenous Australian experiences of Country. Nevertheless, it tries to think, feel and listen to the networked relations and flows between atmospheres, technologies and complex biological systems and views these entanglements 'as' performance. *Performing Climates* aims to unsettle the deeply anthropocentric history of theatre and performance, its obsession with human drama or autobiography, its able-bodied rules and aspirations, its backgrounding of the natural world and its blindness to more-than-human agency. We interrogate what the discipline loses by ignoring the more-than-human and the nonliving. This interrogation builds on the foundational thought of many scholars working in the West who have laboured to unsettle the deeply anthropocentric history of theatre and performance, its obsession with human drama or autobiography, particularly Una Chaudhuri, Wendy Arons and Theresa J. May, Carl Lavery, Bonnie Marranca, Baz Kershaw, Astrida Neimanis, Jennifer Hamilton, Peta Tait, Denise Varney, Lisa Woynarski, Martin Welton, Tanja Beer, Oron Catts and Ionat Zurr.

This book arises in part from Eddie's work organising *Performance Climates*, the 2016 Performance Studies international conference that took place in Naarm (Melbourne). This conference featured the landmark performance *Cut the Sky* by acclaimed Indigenous Australian dance company Marrugeku, keynotes from Richard Frankland, Rebecca Schneider, Peta Tait and Bruno Latour, and resulted in several themed journal editions (Paterson and Brown 2018; Fensham et al. 2018).[5] Our work expands upon this groundswell of voices critiquing Western theatre and performance makings' responses to the climate emergency.

We are working against a model of Western theatre making that might be imagined as 'biospherical'. Recalling the *Spheres* Trilogy from Peter Sloterdijk (2011, 2014, 2016), biospherical theatre sees actions taking place in artificially constructed atmospheres for artistic innovation – proto-laboratories and human greenhouses – in which able-bodied actors are trained and perform within air-conditioned black boxes. Or, more tellingly, white people in white cubes. Biospherical theatre is an enclosed Western form. It is labour intensive, carbon intensive, hierarchical, exclusive, inaccessible and extractive rather than generative of new knowledges and different ways of being with the world. It privileges the human over the more-than-human in all aspects: from theatrical architecture to traditional conservatoire theatre training, to themes, narrative trajectories, design, dramaturgy and direction, casting, consumption, rehearsal and performance style.

In this book, we further dismantle theatre and performance's historical tendency to background the natural world. We develop a model for a more expansive and inclusive theatre ecology. Key to this model are enhanced ideas of theatrical dramaturgy, including Peter Eckersall and Eddie's conception of 'slow dramaturgy' as a means of reconceiving the material conditions of the everyday world in performance such as duration, slowness, place, ambience and bodies both human and nonhuman (Eckersall and Paterson 2011; Eckersall et al. 2017; Eckersall 2024) and Lara's theorisation of 'deep dramaturgy' as an ecological form of dramaturgical practice (Stevens 2019a). Additionally, this book is informed by Angie Abdilla's work developing Country Centred Design; a methodology that centres Indigenous knowledge systems in the development of cultural protocols for the design of new technologies, artificial intelligence and immersive experiences (Abdilla et al. 2021). In *Performing Climates*, these material, ecological and technological modes are also inflected by multiscalar biological and cellular processes of growth, flow and disintegration that exist within a 'planetary performance studies' (Cervera 2017) in which each performance is a cycle with particles and debris (Bleeker and van der Tuin 2014).

Drawing on case studies that might not be necessarily recognisable to Westerners as 'theatre' or 'art', we are interested in performance as a mess of matter, atmospheres, objects, actors and affects: a *compostable art form* that exposes the encounters, transformations and breakdown of cells, of organisms, of bodies, of air, of machines and of light within emergent ecosystems. *Performing Climates* takes into account the cyclical relations and flows between atmospheres, technologies and complex biological systems and views these interactions 'as' performance (Schechner 2002). We see compost in/of performance as a means to reimagine aesthetics and reconnect theatre with climate politics. Our exploration of performing climates necessarily reckons with the climate crisis and its interconnection with the socio-cultural-cosmological fabric of our planet.

Compost as Relationality

The chapters in this book engage with some conventional performances with human actors but are just as interested in what Stacy Alaimo calls 'thinking as the stuff of the world' (2014): the travelling *Agonis* sapling; a melting iceberg; a mermaid who turns into foam; the last breath drawn; the hum of ant air-conditioning; vibrations of voice; the throb of fungal communications; discarded cells of skin; a phallic unicorn horn; underground cathedrals and a Lego-bloom in the Pacific Ocean.

Theatre and performance has always been a political medium, sometimes confirming the status quo and sometimes staging revolt against it. *Performing Climates* brings Western theatre into the century of the climate emergency, to make it a politically 'relevant' and urgent art form again. To make performance-as-usual, which is to say, to churn out the old Western classics without innovation is to perpetuate a culture that has destroyed relationality and kinship between human and the more-than-human world. Performance offers us the tools by which to reframe our relationality – to embody and show others what the world can look like without the human at the centre. It allows us to intermingle with other roots, in ways that might help to radically drive our commitment to protect what is left, to share knowledge of these delicate interconnections of our planetary existence or even to accept the redundancy of our species with grace, courage and humility.

As scholars trained in the rational, ordered, taxonomised Western tradition, with its seemingly pristine lines of demarcation, we are slowly learning to find peace in mess. We do this in order to build on a politics that rejects mastery and those clean binary shapes that obscure the sprawling realities in which we exist. As Alaimo reminds us:

> The anthropocene is no time for transcendent, definitive mappings, transparent knowledge systems, or confident epistemologies. Surely those things got us into this predicament to begin with.
>
> (2016: 3)

Performing Climates thus draws together chapters that are curious about systems of composting matter at different scales and in very different kinds of performances. However, this book does not provide any neat answers to what performance should be or look like as we move through these uncertain times of emergency. All the performances explored in this book are featured because they stage changing states of and sit in tension with biospherical models of theatre.

While writing this book, we have been thinking with the seasons and soil, setting up rat-proof composting bays, navigating the red-bellied black snakes in the vegetable patch, cleaning the gutters, yarning and working with members of the Dhudhuroa Dhargal Aboriginal Corporation to regenerate the land through cultural burning (cool burns), and avoiding the swooping masked-lapwings nesting on the dam. We tell you all this because these are the performing climates which shape our lives and work and because this is a book about the entanglement of systems, weather, particles, organisms, species and art. We agree with 20th century French composer Olivier Messiaen when he described the birds as artists (Odle 2010). We thank Alexis Wright for sharing the butterfly songlines that dance across

Indigenous Country (2023). We invite you to read on, to think compost, to re-picture the world beyond biospherical theatre, to acknowledge place and time and connection and, above all, to enjoy the bloody mess of it all.

Notes

1 Hanna Cormick is of Saami ancestry and has created a performance that grapples with the complex politics of First Nations identification.
2 Examples of collaborative ecological work includes the Indigenous-led projects of Michael-Shawn Fletcher, Jack Pascoe, Uncle Dave Wandin (2022) and the Walbunja Rangers from the Batemans Bay Land Council (Country Needs People n.d.), which draw on Indigenous knowledge and culture to restore the health of Country since invasion and educate Western settlers through traditional fire, land-management and conservation practices.
3 There is no better example of this than the 2024 Venice Biennale Golden Lion award winner Archie Moore's exhibition *kith and kin*. In this expansive work, Moore hand-drew, in white chalk, a 65,000-year First Nations family tree on the black walls and ceiling of the Australia Pavilion. The work included an archive of official redacted documents showing Aboriginal deaths and incarcerations. The chalk drawing included gaping black holes and gaps that point to the violent impact of colonialism on his (and all Aboriginal) ancestry. What was infrequently reported in media coverage of the exhibition and award, however, was that Moore's family tree included nonhuman others as part of the webs of interconnection and kinship lines.
4 Liboiron notes that some First Nations groups refer to Land (with a capital 'L') as an animate entity and ancestor to humans (2021: 7).
5 We acknowledge our colleagues on the organising team *of Performance Climates*: Rachel Fensham, Paul Rae, Robert Walton, Peta Tait, Meredith Rogers, Angharad Wynne-Jones and Alyson Campbell; Melinda Hetzel and the volunteer team; and the many participants of the conference who provided much inspiration for this book. We also thank the contributors, co-editors and teams working on the 'Performance Climates' themed issues of *Global Performance Studies* (Paterson and Brown 2018) and *Performance Research* (Fensham et al. 2018).

References

Abdilla, Angela, Megan Kelleher, Rick Shaw and Tyson Yunkaporta (2021). *Out of the Black Box: Indigenous Protocols for AI. Report*. Melbourne: Deakin University.

Alaimo, Stacy (2014). 'Thinking as the Stuff of the World'. *O-Zone: A Journal of Object-Oriented Studies*. 1(1): 13–21.

Alaimo, Stacy (2016). *Exposed: Environmental Politics and Pleasures in Posthuman Times*. Minneapolis: University of Minnesota Press.

Barad, Karen (2007). *Meeting the Universe Halfway: Quantum Physics and the Entanglement of Matter and Meaning*. Durham, NC: Duke University Press.

Bennett, Jane (2010). *Vibrant Matter: A Political Ecology of Things*. Durham, NC: Duke University Press.

Birch, Tony (2016). 'Climate Change, Recognition and Social Place-Making'. In Eve Vincent and Timothy Neale, eds, *Unstable Relations: Indigenous People and Environmentalism in Contemporary Australia*. Crawley: UWAP Scholarly, pp. 356–383.

Birch, Tony (2019a). 'Walking and Being'. *Meanjin Journal*, Summer 2019, https://meanjin.com.au/essays/walking-and-being/

Birch, Tony (2019b). *HumanNature: Connection and cooperation in a time of climate change.* Australian Museum HumanNature Podcast Series, 14 May 2019. https://australian.museum/blog-archive/museullaneous/humannature-podcast-tonybirch/, accessed 9 October 2024.

Blake, Barry J. and Julie Reid (2002). 'The Dhudhuroa Language of Northeastern Victoria: a Description Based on Historical Sources'. *Aboriginal History*. 26: 177–210.

Bleeker, Maaike and Iris van der Tuin (2014). 'Science in the Performance Stratum: Hunting for Higgs and Nature as Performance'. *International Journal of Performance Arts and Digital Media*. 10(2): 232–245.

Cervera, Felipe (2017). 'Planetary Performance Studies'. *Global Performance Studies*. 1(1). https://doi.org/10.33303/gpsv1n1a3

Country Needs People website (n.d.). https://www.countryneedspeople.org.au/about, accessed 23 April 2024.

Donald, Dwayne (2010). *On What Terms Can We Speak?*. Lecture, University of Lethbridge, Faculty of Education. https://vimeo.com/15264558

Donald, Dwayne (2012). 'Indigenous Métissage: A Decolonizing Research Sensibility'. *International Journal of Qualitative Studies in Education*. 25(5): 533–555.

Eckersall, Peter (2024). *Dramaturgy to Make Visible: The Legacies of New Dramaturgy for Politics and Performance in Our Times*. London: Routledge.

Eckersall, Peter, Helena Grehan and Edward Scheer (2017). *New Media Dramaturgy: Performance, Media and New Materialism*. London and New York: Palgrave MacMillan.

Eckersall, Peter and Eddie Paterson (2011). 'Slow Dramaturgy: Renegotiating Politics and Staging the Everyday'. *Australasian Drama Studies*. (58): 178–192.

Fensham, Rachel, Eddie Paterson and Paul Rae, eds (2018). 'On Climates'. *Themed Issue, Performance Research*. 23(3).

Fletcher, Michael-Shawn et al. (2022). 'Healthy Country'. *Indigenous Knowledge Institute*. https://indigenousknowledge.unimelb.edu.au/Research/themes/2021-2022-themes/healthy-country, accessed 23 April 2024.

Gay'wu Group of Women (2019). *Songspirals: Sharing Women's Wisdom of Country through Songlines*. Sydney: Allen & Unwin.

Graham, Mary (1999). 'Some Thoughts about the Philosophical Underpinnings of Aboriginal Worldviews'. *Worldviews: Environment, Culture, Religion*. 3: 105–118.

Graham, Mary (2009). 'Understanding Human Agency in Terms of Place: A Proposed Aboriginal Research Methodology'. *PAN: Philosophy Activism Nature*. 6: 71–78.

Hamilton, Jennifer Mae and Astrida Neimanis (2018). 'Composting Feminisms and Environmental Humanities'. *Environmental Humanities*. 10(2) November: 501–527.

Haraway, Donna (1988). 'Situated Knowledges: The Science Question in Feminism and the Privilege of Partial Perspective'. *Feminist Studies*. 14(3): 575–599.

Haraway, Donna (2017). 'Symbiogenesis, Sympoiesis, and Art Science Activisms for Staying with the Trouble'. In Anna Lowenhaupt Tsing, Heather Swanson, Elaine Gan and Nils Bubandt, eds, *Arts of Living on a Damaged Planet*. Minneapolis: University of Minnesota Press, pp. M25–M50.

Latour, Bruno (1993). *We Have Never Been Modern*. Cambridge, MA: Harvard University Press.

Liboiron, Max (2021). *Pollution Is Colonialism*. Durham, NC: Duke University Press.

Mathews, Freya (1991). *The Ecological Self*. London and New York: Routledge.

McGrath, Ann and Laura Rademaker (2023). 'The Languages and Temporalities of "Everywhen" in Deep History'. In Ann McGrath, Laura Rademaker and Jakelin Troy, eds, *Everywhen: Australia and the Language of Deep History*. Sydney: UNSW Press, pp. 1–33.

Moreton-Robinson, Aileen (2000). *Talkin' Up to the White Woman: Indigenous Women and Feminism*. Brisbane: The University of Queensland Press.

Moreton-Robinson, Aileen (2015). *The White Possessive: Property, Power, and Indigenous Sovereignty*. Minneapolis: University of Minnesota Press.
Myers, Natasha (2017). 'From the Anthropocene to the Planthroposcene: Designing Gardens for Plant/People Involution'. *History and Anthropology*. 28(3): 297–301.
Neale, Margaret (2022). 'First Knowledges: An Introduction'. In Zena Cumpston, Michael-Shawn Fletcher and Lesley Head, eds, *Plants: Past, Present and Future*. Melbourne: Thames and Hudson, pp. 1–4.
Odle, Melissa Patrice (2010). 'Bird Song in the Music of Olivier Messiaen, With an Emphasis on *Le Merle Noir*'. Master of Music in Performance Thesis, California State University, Northridge.
Paterson, Eddie and Kevin Brown, eds (2018). 'Performance Climates'. Themed Issue. *Global Performance Studies*. 1(2). https://gps.psi-web.org/issue/view/2
Plumwood, Val (1993). *Feminism and the Mastery of Nature*. London and New York: Routledge.
Rose, Deborah Bird (1992). *Dingo Makes Us Human: Life and Land in an Australian Aboriginal Culture*. Cambridge: Cambridge University Press.
Rose, Deborah Bird (2017). 'Shimmer: When All You Love Is Being Trashed'. In Anna Tsing, Heather Swanson, Elaine Gan and Nils Bubandt, eds, *Arts of Living on a Damaged Planet: Ghosts and Monsters of the Anthropocene*. Minneapolis: University of Minnesota Press, pp. 51–63.
Salleh, Ariel (1997). *Ecofeminism as Politics: Nature, Marx and the Postmodern*. London: Zed Books.
Schechner, Richard (2002). *Performance Studies: An Introduction*. London: Routledge.
Sloterdijk, Peter (2011). *Bubbles: Spheres Volume I: Microspherology*, trans. Wieland Hoban. South Pasadena CA: Semiotext(e).
Sloterdijk, Peter (2014). *Globes: Spheres Volume II: Macrospherology*, trans. Wieland Hoban. South Pasadena CA: Semiotext(e).
Sloterdijk, Peter (2016). *Foams: Spheres Volume III: Plural Spherology*, trans. Wieland Hoban. South Pasadena CA: Semiotext(e).
Stengers, Isabelle (2018). *Another Science Is Possible: A Manifesto for Slow Science*, trans. Steven Muecke. Cambridge: Polity Press.
Stevens, Lara (2019a). 'Anthroposcenic Performance and the Need for "Deep Dramaturgy"'. *Performance Research*. 24(8): 89–97.
TallBear, Kim (2014). 'Standing with and Speaking as Faith: A Feminist-Indigenous Approach to Inquiry'. *Journal of Research Practice*. 10(2): 1–7.
Tsing, Anna Lowenhaupt (2015). *The Mushroom at the End of the World: On the Possibility of Life in Capitalist Ruins*. Princeton, NJ: Princeton University Press.
Watson, Irene (2014). 'Re-Centring First Nations Knowledge and Places in a Terra Nullius Space'. *AlterNative: An International Journal of Indigenous Peoples*. 10(5): 508–520.
Wilson, Alex (2008). '"N'tacinowin Inna Nah": Our Coming in Stories'. *Canadian Woman Studies*. 26(3): 193–199.
Wilson, Shawn (2008). *Research Is Ceremony: Indigenous Research Methods*. Black Point, Nova Scotia: Fernwood Publishing.
Wright, Alexis (2023). *Praiseworthy*. Sydney: Giramondo.

2 Biospheres

> A high-resolution aphrogram of a "society" would render visible the honeycomb and neighbourhood system of air-conditioned bubbles, thus making it clear that "societies" are polyspherical air-conditioning systems, in both the physical and psychological sense ... From that point on, the political field would have to be examined with the aid of a fluid mechanics for semantic changes or vectors of sense.
>
> (Sloterdijk 2016: 281)

Let us begin with a series of spheres. In this case, the spheres belong to water molecules that are part of water vapour. The vapour is about to move through the vast air-conditioning system that modulates the temperature and atmosphere of the Flower Dome biome at Singapore's Gardens by the Bay (Bellew and Davey 2012: 74–81). The vapour is part of the fresh air that is drawn into the system from outside of the dome. The atmosphere in Singapore is hot and humid. The air is sucked into the dome through an intake louvre and the water vapour travels into a desiccant conditioner. This process sees the vapour pass through a salty liquid, as a means of removing moisture from the air. The dehumidified air then reaches an air handling plant room. The water vapour in the air has been reduced, though some still remains. The air passes through ventilation points in the walkways, walls and planting areas. The water vapour enters the dome and is pushed through the space by mounted fans that ensure a regular air velocity.

A water vapour molecule encounters the leaves of *Agonis flexuosa* (commonly known as the Willow Myrtle, the Western Australian Peppermint or the Swan River Peppermint, yes, the same peppermint as the one that stands outside our home) in the Australian Garden. Some of the vapour condenses on the surface of a leaf, it forms a minute droplet which slips down the leaf and lands on the soil. The roots of the *Agonis* sense the moisture and move to take it in. Soon the water molecule is part of a plant body, combined with carbon dioxide it becomes glucose and oxygen. The oxygen is released through the cells of the tree as it respires. As the tree respires, the oxygen interacts with the atmosphere inside the dome, revealing networks of organic

DOI: 10.4324/9781003273080-2

and inorganic matter, technologies of metal, glass and soil, the human and more-than-human. It might be said that the vapour exists within a series of intersecting spheres.

In his sphere trilogy (*Bubbles*, *Globes* and *Foams*), German philosopher Peter Sloterdijk situates the sphere as the 'world shaping' conceptual frame that defines the parameters of modern human existence from the microsphere of the bubble and womb to the macrosphere of the globe. In this Eurocentric conception, Sloterdijk deploys an image of Shakespeare's Globe Theatre in London to show how a spherical vision of globalisation – globes, greenhouses and domes – reveals the shaping of Western nations, states, climates, economies, politics, societies and culture since the Enlightenment (2014: 877). We are drawn to the biodomes of Singapore's Gardens by the Bay and to Shakespeare's Globe Theatre, as these two spherical spaces offer strangely interlinked examples of how particular kinds of performing climates have become dangerously embedded into our contemporary practices of thinking and making theatre and performance.

Expanding on Sloterdijk, we position the image of the Globe Theatre as the modern archetype for Eurocentric and Western approaches to making and thinking about performance.[1] The Globe can be understood as a kind of biome: a material, metaphorical and psychological architecture that has helped determine how performance as an art form, concept and way of thinking has evolved through modernity. In Sloterdijk's terms, a biome is a 'capsule delirium' that so obsessively controls its atmospheric and biospheric dimensions it takes on the 'anticipatory' shape of a dire warning about our anthropotechnological everyday (2016: 332). The architecture of the Globe is inward-looking and centre-focused. The image of the Globe Theatre as a sphere – a kind of greenhouse of modernity and a proto-laboratory for performance in which humans gather to look at themselves – suggests an insular, navel-gazing space for artistic innovation, experimentation and entertainment.

The image of the Globe also allows us to reconceptualise Western practices for making and thinking about theatre and performance as a series of spheres that have been evolving through modernity but always with the same embedded architecture. Inside these Globes, art develops in enclosed and air-conditioned bubbles (laboratories, rehearsal rooms, conservatories and galleries). We conceptualise this kind of theatre as being biospherical: a human-centric endeavour, evolving inside a globe, largely upholding the fantasy of itself as disconnected from atmospheric and environmental interactions beyond the human.

The curvature of the planet as a globe is most evident when we travel by air from our home in Australia. As employees of Western universities, working in the fields of performance and theatre studies, our travel for conferences and fieldwork has often meant travelling to Western centres of knowledge and culture in the United Kingdom, Europe and the United States. Flying from Australia is carbon intensive and takes a long time, it also frequently involves crossing over, or stopping off at, the tropics of South East Asia. For us, for a time (like many others we have been trying to reduce our travel over the past decade), overseas travel meant disembarking into the humidity of Singapore and visiting family and friends who happened to be based there with repeated visits to Singapore's Gardens by the Bay complex and Flower Dome.[2]

In 2012, Singapore opened Gardens by the Bay, an ecological park created by the Singapore National Parks Board on 101 hectares of reclaimed land, located behind the towering Marina Bay Sands Casino and adjacent to the financial district. Approaching from a distance the first thing we notice are the spectacular 'supertrees', giant wormhole-shaped lattices constructed out of concrete and metal that serve as elaborate vertical gardens, stretching 25–50 metres high. The Gardens also feature two specially designed glass domes: one that houses a flower garden and the other a 'cloud forest'. These dramatic structures are best described as reverse greenhouses: just as Europeans developed greenhouses to enable the growth of warm-temperature plants in cold environments, the domes are climate controlled to simulate cold temperatures that allow temperate-zone plants to flourish in the steamy surrounds (Bellew and Davey 2012).

Indeed, Lee Kuan Yew, the founder and long-time prime minister of Singapore, once cited the air-conditioner as the most influential invention of the last millennium. All the world's pre-eminent civilisations, he said, had developed in cooler climates (McCarthy 1999; see also George 2000). For humans, rudimentary systems of cooling have long been important because of their capacity to slow bacterial growth in water molecules – and thereby slow the rate of organic decomposition in food and the growth of bacterial infections.

One of the earliest 'modern' examples of air-conditioning is likely American physician and scientist John B. Gorrie's ice machine of 1844. For Gorrie, the refrigeration of air and the formation of ice was explicitly linked to the formation of atmosphere and his machine can be considered a forerunner of the air-conditioner (Nagengast 2009). Gorrie advocated installing cool airflows in hospitals as good practice for treating patients and reducing the rates of diseases such as malaria and wrote ambitious plans to cool entire cities in the Southern United States (Roth 1936). Ice slows time. Melting ice changes atmospheres and air-conditioning changes social spheres.

For Yuriko Furuhata, climatic control is at the basis of an atmospheric future in which weather forecasting and air-conditioning technologies will become more precise, customised and individualised (2022: 50–51). Within the air-conditioned spaces of cars, offices, homes and theatres, we already each exist in our own compartmentalised microclimates. The image of the biodome, then, is not so much a technological innovation – as seen in the 'Aerocene' experimental architectures of Tomás Saraceno, or the gigantic bubble covering New York City from Buckminster Fuller and Shoji Sadao's 'Dome over Manhattan' (1960) – but a climate controlled structure that echoes our atmospheric reality (Furuhata 2022). We live in an age of biodomes.

In Singapore's humid climate, air-conditioning is not only linked to practices of so-called civility and health but also strongly linked to the nation's wealth and economic growth. Artificially controlled climates allow human workers to sustain longer working days in temperatures that are more comfortable for the fragile human scales of variability. Thus, air-conditioning Singapore is both part of the colonial 'civilising' project in Asia, key to its economic advancement and, through projects like Gardens by the Bay, an example of biotechnological innovation at the intersection between power, politics, capital, wealth, ecology and climate control.

Gardens by the Bay achieves Lee's vision but, rather than housing humans, it supports flora and fauna (Figure 2.1). In fact, in the initial designs, even the most sophisticated air-conditioning system was inadequate as the direct Singaporean sun on the glass domes made it impossible to cool the inside of the spaces, forcing architects and engineers to invent new ways of cooling the air. These design and engineering solutions include the development of a special glass that lets in light (for plant growth and photosynthesis) but not heat; shade sails that automatically protect the plants from too much sun; systems to dehumidify the air before cooling; and thermal stratification achieved by cooling the lower zones and venting hot air at the top (Bellew and Davey 2012).

The Gardens' promotional and educational materials emphasise that these cooling technologies reduce energy consumption and that such innovations are part of a broader strategy of sustainable design and ecology. The supertrees have solar cells for power and serve as heat exhausts for the domes, and the gardens as a whole generate 8% of their own electricity using steam generated from decaying plant waste.

With its artificial engineering, technological and scientific innovations, we submit that the Flower Dome is a theatrical space. It organises the dramaturgical elements of light, flow, sound, time, body and atmosphere in an enclosed area in which the conditions of performance are easily recognisable. There is an audience, performers, Muzak piped through the speakers in the ceiling and a tightly monitored environmental architecture.

As we consider the Flower Dome, we take into account the interaction between biological and mechanical agents, and human and more-than-human actors. The performance here, of plants, of bodies, of air, of machines, and of light, takes place in an enclosed dramaturgical ecosystem. As distinct from models of 'Tropical Futurity' that are embedded in the particular temporal, climatic and socio-historical conditions of South Asian countries situated close to Singapore, such as the Philippines (Carreon 2024), the biospheres of Gardens by the Bay are Eurocentric and artificial by design and conception (Myers 2015). The Flower Dome offers a cross-section image of a laboratory in which experimentation, labour, creation, capital and repetition are embedded within a sphere. It many ways, it operates as a contemporary version of Shakespeare's Globe. It is an image of biospherical theatre.

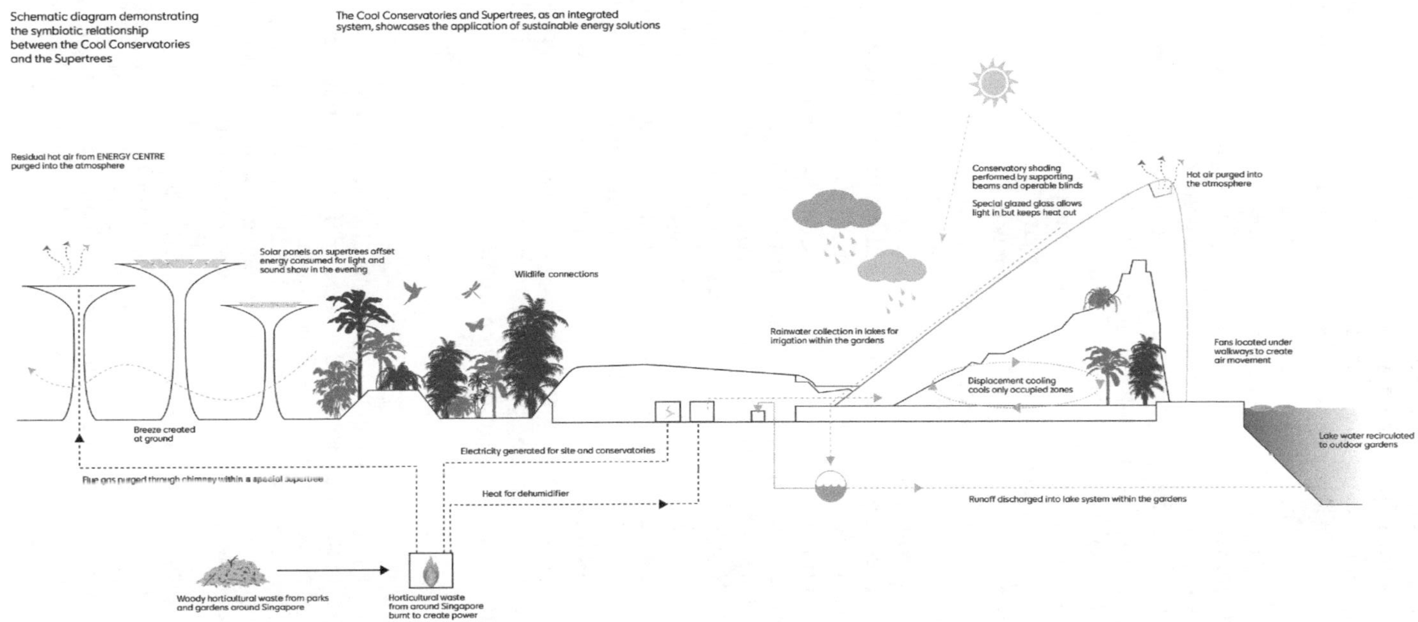

Figure 2.1 Gardens by the Bay plan.

Yet, alongside systems of enclosure and air-conditioning, muck and mess are part of this performing climate. Crucially, in the case of the Flower Dome, there is leakage and waste, and such processes challenge the existing conditions and offer new theatrical possibilities. As part of the design of the Flower Dome biome, the artificially air-conditioned atmosphere must engage with the louvres and doors. These airflows facilitate the breakdown of large amounts of biomass which are recycled from the gardens, parks and streets of Singapore. This organic waste is used to power boilers, becoming ash that is applied as high-grade fertiliser to the plants and supplying electricity to the chillers and desiccant regenerator that are integral to the air-conditioning system of Gardens by the Bay (Bellew and Davey 2012: 84–91). Indeed, it is the elements of waste – bacteria, debris and breakdown – that pierce what appears as the bubble of the biome, bringing in other more-than-human interconnections.

The cyclical action of breakdown and renewal adds a counter-dimension to this familiar biospherical mode and a vision of performance that moves outside of the globe. This means that the movement of vapour cannot be simply recycled through an enclosed atmosphere but is also part of a larger network of bacteria, fungi and microbes embedded in the soil and interacting with the cellular networks of trees and plants. The dramaturgy of processes of exchange and transformation in the *Agonis*, for example, invites us to see how all living processes move across and through biological, social, political and climate spheres – a movement towards *compost*, the interaction of living and nonliving elements in a cycle of decomposition and re-composition. This cycle involves the breakdown of biological waste and matter, messy entanglements with bacteria, protozoa, nematodes and fungi that fundamentally change the physical, chemical and biological properties of this artificially constructed climate. This act of *composting* reveals how, even when overwhelmingly framed by biospherical conditions, natural processes can teach us new ways of thinking about performance if we are attentive to the intricacies of more-than-human choreographies.

Notes

1 As Marc Jongen notes: 'Like all illustrations in Sloterdijk's books, it is not merely an illustration of the ideas developed in the text. These illustrations are designed to plant in the reader's mind an awareness of the figurative, encouraging them to "see" or to "imagine" what is written' (2011: 217). In drawing attention to the image as thought – a way of seeing or imagining – Sloterdijk suggests that our thinking about the world-as-image is also a way of navigating contemporary life and, potentially, reimagining it (Jongen 2011: 199). For an extended analysis of Shakespeare's Globe as an image of modernity, see Wilson (2015).

2 Initial thinking on Gardens by the Bay as a site of performance was co-authored with Hallam Stevens and featured in a 2014 paper at the International Federation of Theatre Conference entitled 'Performing the Future: Singapore's Biopolis'. We thank Lara's brother for his contributions and for introducing us to Singapore and Gardens by the Bay.

References

Bellew, Patrick and Meredith Davey (2012). *Green House, Green Engineering: Environmental Design at Gardens by the Bay*. Novato, CA: Oro Editions.

Carreon, Erika (2024). 'Tropical Futurity: Speculative Ecofiction through the Lens Tropical Temporality'. Ph.D. Thesis, University of Melbourne.

Furuhata, Yuriko (2022). *Climatic Media: Transpacific Experiments in Atmospheric Control*. Durham and London: Duke University Press.

George, Cherian (2000). *Singapore, the Air-Conditioned Nation: Essays on the Politics of Comfort and Control, 1990–2000*. Singapore: Landmark Books.

Jongen, Marc (2011). 'On Anthropospheres and Aphrogrammes: Peter Sloterdijk's Thought Images of the Monstrous'. *Humana.Mente Journal of Philosophical Studies*. 18: 199–219.

McCarthy, Terry (1999). 'Lee Kuan Yew'. *Time Magazine*, 23 August. http://content.time.com/time/world/article/0,8599,2054444,00.html#ixzz2lzv7Abbn, accessed 29 November 2013.

Myers, Natasha (2015). 'Edenic Apocalypse: Singapore's End-of-Time Botanical Tourism'. In Heather Davis and Etienne Turpin, eds, *Art in the Anthropocene: Encounters among Aesthetics, Politics, Environments and Epistemologies*. London: Open Humanities Press, pp. 31–42.

Nagengast, Bernard A (2009). 'Was Ice Making John Gorrie's Greatest Legacy?'. *ASHRAE Transactions*. 115(1): 122–129.

Roth, George B (1936). 'Dr. John Gorrie: Inventor of Artificial Ice and Mechanical Refrigeration'. *The Scientific Monthly*. 42(5): 464–469.

Sloterdijk, Peter (2014). *Globes: Spheres II*, trans. Wieland Hoban. South Pasadena CA: Semiotext(e).

Sloterdijk, Peter (2016). *Foams: Spheres Volume III: Plural Spherology*, trans. Wieland Hoban. South Pasadena CA: Semiotext(e).

Wilson, Richard (2015). 'A Globe of Sinful Continents: Shakespeare Thinks the World'. In Jernej Habjan and Fabienne Imlinger, eds, *Globalizing Literary Genres: Literature, History, Modernity*. New York: Routledge, pp. 27–49.

3 Death

Let us begin with death. Death is a perfectly logical place to start when you understand death not as an endpoint nor in the monotheistic religious sense of eternal life after earthly death, but rather as a persistently active composting transformation of the old into new life. The structure of this book is also compostable, which is to say it does not necessarily have a beginning and end, but instead it moves in flows, cycles and repetitions (which means you can dip into it anywhere you like!). It anticipates its own composting – its pages one day fuel for the fire, its digital ghost haunting the recesses of an overcrowded Internet. In talking about death and more-than-human actors, it is not because we wish to kill theatre. On the contrary, theatre, it is because we love you and all your vibrant, dizzying, spark that we celebrate your death, your decomposition and re-composition before our eyes.

In Hatha Yoga texts dating back to the fifteenth century in South Asia, *Savasana* is the 'corpse pose'. It is practised by the yogi as a rehearsal of the state in which we will all end up one day. It is the last pose in the yoga class, the one that everyone 'looks forward to' but is, in fact, the hardest to perform – the act of letting go. Lying prone on the floor, we perform cadaver. We take our mind's eye to our bones and invite them to sink down heavily into the mat, the parquet, the earth beneath. We watch the rise and fall of our chest, the cold intake of air against our nostrils and the inside cavities of the nasal passages. We watch the movement of air filled with oxygen and water vapour as it moves down our respiratory tract and into our lungs where it changes places with the carbon dioxide which gets carried out of the body and exhaled. Or perhaps we watch the microplastic bead that we inhaled, sucked into our lungs and lodged there, indefinitely.

In *Savasana*, we have more opportunities to be attuned to the dead skin cells on our arms falling to the floor, making way for new ones to grow beneath. Our hair falls from our scalp and our follicles shorten. We summon our consciousness, our spirit if you like, to come out of our body and float to the ceiling, watching from above this dead weight, this lump of composting flesh lying below us. In this compostable state, the ego is forced to let go – there are no achievements to be realised, beauty becomes irrelevant, self-consciousness is stripped away in the lonely act of dying and watching

DOI: 10.4324/9781003273080-3

yourself dying. It is too late to regret what we have done or look forward to what is to come, we can only be in the here and now.

In Western culture, theatre hosts a form of *Savasana*. It is a place for playing dead, for rehearsing or practising, without anxiety, that part of the cycle of life. In Forced Entertainment's show *Spectacular* (2008), a figure of death stands on the stage in a Halloween skeleton costume and describes the scene for us while also dying on stage, again and again. Tim Etchells (2008) notes the strange in-between-ness that theatre creates in such moments of on-stage dying:

> No-one's fooled. No-one's taken in. Doesn't matter how much fake blood, how much yelling, how much sobbing, how much stillness. No one thinks this is real. But at the same time there remains a strange charge to this game, a cultural and emotional electricity which crackles and sparks the air around the actor who lives-but-dies, or who lives but plays dead.

What charges that frisson of watching living bodies in states of fake dying? Is it the liminal space of theatre, positioned between life and death, real and not real, growth and compost that makes it a temple of spiritual return for the West? Etchells points out that death is part of the Bloody Mess of life and theatre persistently revisits death in forms that our fragile egos can tolerate – in the ghost of Hamlet's father, Godot's nonappearance or *Spectacular*'s grim reaper. In theatre, this (usually) live event populated with living creatures, we always come back to death (see Carlson 2003). *Spectacular*'s parody of this theatrical paradox suggests to us that theatre might hold a key to unsettling this binary and wake us up to the composting that is occurring before our very eyes on stage, at all times.

And what of the French philosopher Hélène Cixous, who questioned whether she could attend any more theatre in 1977 because all the productions she saw were committing femicide. As she remarked, '[i]t is always necessary for a woman to die in order for the play to begin' (1984: 546). As ecofeminists, we too turn our back on representations of death as plot devices and we see the Freudian death drive as part of a mentality that justifies masculine projects of mastery which have killed or damaged so much life on this planet (Plumwood 1993). We see death in compost theatre as radically different to this kind of theatricalisation of murdering women and girls to up the stakes (and the melodrama) in a narrative. Since only in Western philosophy are the states of life and death turned into binaries, we are interested in performances where death is life-giving and life-affirming. As philosopher Val Plumwood put it, when we think about dead bodies feeding the organisms and microorganisms in the earth, death can 'affirm an enduring, resilient cycle or process' (2007a: 67).

Death is necessarily bound to time. In many Indigenous cultures and languages, conceptions of time and ontology are not flat, linear historical chronologies that we, in the West, are so familiar with but instead are configured as webs of connection (McGrath and Rademaker 2023: 6). Australian Indigenous cultures, the oldest living cultures on the planet, experience time through ancestorial connections with Country. White anthropologist W.E.H. Stanner usefully described/translated these temporal understandings as the 'everywhen'. The everywhen emphasises the ontological entanglement of time-place-space and even aural performance. This encompasses materiality, embodiment and language, assuming a nonseparation between mind, body, spirit, name, totem, word/song/movement and features of landscape as animate storyteller in its own right (McGrath and Rademaker 2023: 14). Such a holistic philosophy interlaces the human in a relationship of obligation and reciprocity with earth others, inviting past ancestors and future relatives into the present and situating them within particular places and features of Country. Such a complex space-time conception that activates a living history necessarily contrasts with Western human understandings and practices of mourning for both the human and the more-than-human.

Australian environmental philosopher Val Plumwood was heavily influenced by Indigenous notions of death, place, time and the position of the human in the natural order (see Plumwood 2007b). The performances discussed in this chapter invite such a radical rethinking of death for Western audiences, revealing where dead organic matter turns over into fresh, vitalising substances. These compost performances are interested in the spectral qualities of living, while exploring the question of how our species might die with humility and what role theatre can play in that process.

Humans as Food

In 2023, Belgian visual artist and theatremaker Kris Verdonck, in close collaboration with the dramaturg Kristof van Baarle, created an avant-garde music-theatre performance *PREY*. *PREY* is based on the life and work of pioneering philosopher Val Plumwood.[1] Plumwood died from a stroke in 2008 but, during her lifetime, wrote extensively about death, decomposition and Western burial practices. Verdonck's company has long been interested in staging death and is well known for creating sparse, post-apocalyptic stages with circular, Beckettian monologues and polylogues. The desolate titles of some of his works give a sense of his thematic interest in representations without human actors or where the human is rendered less interesting or more vulnerable than machines (see Grehan 2015), including *IN VOID* (2016), *CONVERSATIONS (at the end of the world)* (2017), and *SOMETHING (out of nothing)* (2019). Van Baarle describes these works as interrogating the spectral remains of humans in the theatre, staging a post-theatre or after-theatre (2019: 228).

Plumwood's story is a ready-made narrative of human composting. In 1985, Plumwood narrowly survived an attack by a saltwater crocodile which profoundly influenced the direction of her environmental philosophy.[2] Paddling solo in her canoe in the mangroves of Kakadu National Park, in the

Northern Territory, a crocodile sprung out of the water, put its huge jaws around her thigh and wrenched her from the boat and into the water (2012: 11). The crocodile pulled Plumwood into a death roll three times, a common method for drowning prey before consuming it. Eventually it gave up on the stubborn meal and let her go. Bleeding profusely from wounds on her legs and torso, Plumwood crawled up the bank and waited, dying in the long grass. As she waited to be discovered, which she eventually was by a park ranger, she describes becoming radically and newly self-aware as 'being part of the feast and not just some sort of spectator of it' (2012: 15).

Verdonck's *PREY* is a three-part work, played by three different female performers of different generations and based on the structure of a Japanese Noh play. The first part is a monologue, the second an avant-garde operatic performance and the third a dance piece. Veteran Belgian theatre actor Katelijne Damen delivers the monologue as a narration in third person. *PREY* begins with a haunting avant-garde soundscape and Damen lying, fully clothed, like a cadaver on the stage next to a ream of red fabric. When she eventually moves and rises from the floor, she begins a monologue about the injuries resulting from the crocodile attack. Plumwood, a great philosophical mind, is reduced to flesh and blood. She is not only food for the crocodile, even her 'vomit [is] a feast for fish and insects' in the vast mangroves and swamps of the park (Verdonck 2023). *PREY* theatricalises Plumwood's radical repositioning of her place in the world and the way it violently forces her into humility. As food for the crocodile, the human is no longer exceptional once confronted with her true place in the natural order of the planet and the mutual dependence upon all species in a thriving ecosystem. For Verdonck and van Baarle, whose work assumes the immanent collapse of our species in a climate emergency, the work captures the experience of living in 'end times' (van Baarle 2020).

Since Noh plays are ghost plays, Damen's body suggests a haunting, the materiality of an older woman, roughly the age that Plumwood would have been at the time of the performance had she still been alive. She describes Plumwood as a combination of sugars, fats and carbohydrates: '[h]er body, a meal not to be missed... She was the banquet. Alive or dead. Moving or rotting' (Verdonck 2023). She links the human place within the food chain as prey and predator to the environmental crisis more broadly: 'If we are not part of our body, we are not part of our food, we are not part of the earth we walk on' (Verdonck 2023). The implication is that if we cannot even see our real place within the natural order, how will we ever learn to treat the earth as anything more than an inexhaustible servant to our needs and whims?

In part two, Anna Clare Hauf sings the libretto of an experimental opera. The performance is overshadowed, literally and figuratively, by a piece of material or curtain draped in a cocoon-like formation. It sits centre-stage and slowly grows throughout the scene, eventually towering over the singer, the musical ensemble and the audience like a creature from the Upside-Down. The performance draws from Plumwood's interest in Indigenous

Figure 3.1 Mooni Van Tichel as the crocodile in *PREY*, Kris Verdonck, 2023.

philosophies and creatures that shapeshift between worlds, moving between human and more-than-human. In particular, *PREY* references Plumwood's description of the Indigenous world creator called 'Ngalyod'. For the Kuninjku people of Western Arnhem Land in the far north of Australia, Ngalyod is a rainbow serpent in a creation story, a figure that connects everything in the universe together and moves between human and nonhuman forms (2012: 9). *PREY* also references Plumwood's account of other Kuninjku beings, the Yawkyawk sisters, Likanaya and Marrayka, freshwater spirits similar to mermaids but which can also take human forms (2012: 26–27).

The final scene, which featured the youngest performer, dancer Mooni Van Tichel, embodied the fluidity between human and more-than-human (Figure 3.1). Van Tichel, wearing a scaled bodysuit, switched between performing the crocodile hunting for her meal and the human meal thrashing against the attack. Peter Eckersall writes, 'her flesh seems to leak out of the slits in the costume, that is reptilian with hints of textured landscapes into which the human is submerged' and notes that there is a struggle for the human as she wrestles with becoming animal throughout the dance (Eckersall 2024: 125). Van Tichel rolls across the stage, diving on her prey. The next moment her movement slows and gives the impression that she is being pulled under the water, kicking in silence and slow motion, writhing against an unseen attacker. Yet, *PREY* is not a work of redemption or even survival or the triumph of the human over a predator. It is a work that invites a new perspective on the human as an 'energetic feast of digestible meat' in 'a story with no beginning and no end' (Verdonck 2023). A work of compost theatre.

Farewell

The question of human death and our place within the natural order is explored in a radically different earlier Japanese performance *Sayonara* (2011). In March 2011, an earthquake in Fukushima, off the coast of Japan, caused a tsunami which precipitated the meltdown of reactors at the Fukushima nuclear power plant, creating the worst nuclear disaster since Chernobyl in 1986. Five months after the nuclear disaster, Lara travelled to Osaka to attend a conference where she saw the play *Sayonara*, written and directed by Oriza Hirata.[3] *Sayonara* is a play about a young woman, played by the Canadian actor Bryerly Long, whose character has a terminal illness.[4] When the lights come up on the stage, it is not Long's character the spectators are first introduced to but instead a beautiful young Japanese woman sitting in a chair. This character, it is later revealed, is a companion for the young woman, hired by her father to comfort his daughter through her last days. As the lights became brighter on the stage, it became increasingly apparent to spectators that something was not quite right about the caregiver. It turned out that the caregiver was not performed by a human but instead an android, Geminoid F, created by roboticist Hiroshi Ishiguro. Geminoid F performs as 'herself', as robot trained to perform as an empathetic nurse, reciting poems and engaging in respectful conversation to comfort the young woman.[5]

Sayonara has received much scholarly and critical attention for its cutting-edge technological innovation and the questions this raises about the potential redundancy of human actors (see D'Cruz 2014; Spedalieri 2014; Eckersall 2015).[6] We are interested in the performance for its portrayal of death and nuclear contamination. When viewing the performance in 2011, it was impossible not to contextualise the work in relation to the Fukushima disaster. Though the performance never names the woman's illness, given the history of Japan in World War II as well as the recent disaster, the first thought is that she is dying from radiation poisoning. Indeed, in a later version of the performance that we saw in Melbourne in 2012, Hirata added a coda to the original piece in which the young woman has passed away. Here, the android is assigned a new role: to go to the exclusion zones, now saturated and uninhabitable with radiation, to recite poems to the victims of the fallout. This echoed the work of actual robots following the catastrophe which were sent into the Fukushima fallout zone to attempt to clean up the spill and check the on-site reactors at the plant. These robots have since been declared to have 'died' after failing their missions, damaged by radiation or not sophisticated enough in their engineering to penetrate the complex terrain of the site (Crew 2016; McCurry 2017).

The post-catastrophe and posthuman scenario of *Sayonara* challenges the viability of human futures within an environment contaminated with radiation for thousands of years. The scenario demands thinking in the everywhen, compressing past and future into a laterally expanding present. Despite being staged in a black box, the posthuman future that *Sayonara* depicts powerfully evokes what is absent – the poisoned soil, waterways and species and all the future generations impacted by the contamination. What is most post-apocalyptic is its depiction of machines trained to perform emotions and acts for which there can never be a genuine, felt experience behind – the manufacturing of human care and empathy, as if it were another commodity for techno capitalism to produce and sell. The play does not dwell on the woman's physical suffering so much as her loneliness, and the human character seems to stoically accept the performance of the android's sincerity with disturbing acquiescence. The tragedy is that there is never any sense that there might be a human alternative; it is too late. What Plumwood calls 'human chauvinism' is displaced by this dire and unusual but not entirely unrealistic scenario (2012: 71).

Métis scholar Michelle Murphy notes the entanglements of the human with chemical compounds, the imbrication of our bodies with those poisons we have put on the planet and their uneven impact on different bodies (2018). She calls this 'alterlife', 'a figuration of chemical exposures that attempts to be as much about figuring life and responsibilities beyond the individualized body as it is about acknowledging extensive chemical relations' (2017: 497). Murphy positions it as a major product of human capitalist and colonial relations that treat the earth as a dumping ground but also points out that chemical relations long preceded human life and will continue to influence the future of our species. For Murphy:

> Alterlife acknowledges that one cannot simply get out, that this hurtful and deadly entanglement forms part of contemporary existence in this moment, in the ongoing aftermath. And yet the openness to alteration may also describe the potential to become something else, to defend and persist, to recompose relations to water and land, to become alterwise in the aftermath.
>
> (2017: 500)

As we bid farewell to the frail human body in *Sayonara*, we are invited into different timescales of substances that outlive both the human and the robot. We in the West, and highly industrialised nations such as Japan, are only now, almost a century after the development of atomic energy, starting to better understand these webs of connection, recognising our entanglement with the world, accepting these states of becoming and maybe even learning from them in our alter-wisdom.

Letting It All Go

Performance-making in the time of the sixth mass extinction invites deep reflection on the role of theatre as a place where humans have long rehearsed the difficult act of dying and our ghostly afterlives. In a time moment when human and more-than-human existence are under threat from industrial practices, compost theatre invites a humbling reflection on our past actions and attitudes, our present place within the food chain and the future viability of our species on a contaminated planet. Through their depiction of the more-than-human, both organic and technological, these performances radically decentre the human as central to their ecological consciousness. These performing climates show us alternatives or alterlives, a promise that, with or without us, the vibrant exchanges of life and death on this planet will persist.

Notes

1 Lara sits on the board of Plumwood's rainforest property and former house, now a living archive, in southern New South Wales. Lara helped coordinate Verdonck and van Baarle's visit to Plumwood Mountain which formed part of their research into the life and experiences of Val Plumwood.
2 Plumwood first described her experience in the essay 'Human vulnerability and the experience of being prey' (1995).
3 We both saw a later iteration of *Sayonara* on 25 August 2012 at The Arts Centre, Melbourne.
4 The actor is a blonde, Caucasian woman.
5 The absent family in the face of the young person's tragedy recalls Japan's ageing population and the need for alternative caregiving solutions for the baby boomer generation.
6 D'Cruz (2014) notes that the stereotypical racial and gendered features of the robot are critical to its believability as a caregiver.

References

Carlson, Marvin (2003). *The Haunted Stage: The Theatre as Memory Machine*. Ann Arbor: University of Michigan Press.
Cixous, Hélène (1984). 'Aller à la mer', trans. Barbara Kerslake. *Modern Drama*. 27(4): 546–548.
Crew, Bec (2016). 'The Robots Sent into Fukushima Have "Died"'. *Science Alert*, 11 March. www.sciencealert.com/the-robots-sent-into-fukushima-have-died, accessed 7 November 2023.
D'Cruz, Glenn (2014). 'Things I Know About Geminoid F, or What I Think About When I Think About Android Theatre'. *Australasian Drama Studies*. (65): 272–288.
Eckersall, Peter (2015). 'Towards a Dramaturgy of Robots and Object-Figures'. *TDR: The Drama Review*. 59(3): 123–131.
Eckersall, Peter (2024). *Dramaturgy to Make Visible: The Legacies of New Dramaturgy for Politics and Performance in Our Times*. London: Routledge.
Etchells, Tim (2008). '"Spectacular" Programme Note'. *Forced Entertainment*, www.forcedentertainment.com/spectacular-programme-note-by-tim-etchells, accessed 7 December 2023.
Forced Entertainment (2008). *Spectacular*. Performance Work.
Grehan, Helena (2015). 'Actors, Spectators, and "Vibrant" Objects: Kris Verdonck's *ACTOR #1*'. *TDR: The Drama Review*. 59(3): 132–139.
Hirata, Oriza (2011). *Sayonara*. Performance Work, Seinendan Theatre Company.
McCurry, Justin (2017). 'Fukushima Nuclear Cleanup Falters Six Years after Tsunami'. *The Guardian*, 9 March. www.theguardian.com/world/2017/mar/09/fukushima-nuclear-cleanup-falters-six-years-after-tsunami, accessed 1 September 2023.

McGrath, Ann and Laura Rademaker (2023). 'The Languages and Temporalities of "Everywhen" in Deep History'. In Ann McGrath, Laura Rademaker and Jakelin Troy, eds, *Everywhen: Australia and the Language of Deep History*. Sydney: UNSW Press, pp. 1–36.
Murphy, Michelle (2017). 'Alterlife and Decolonial Chemical Relations'. *Cultural Anthropology*. 32(4): 494–503.
Murphy, Michelle (2018). 'Against Population, towards Alterlife'. In Adele E. Clarke and Donna Haraway, eds, *Making Kin Not Population*. Chicago: Prickly Paradigm Press, pp. 101–124.
Plumwood, Val (1993). *Feminism and the Mastery of Nature*. London: Routledge.
Plumwood, Val (1995). 'Human Vulnerability and the Experience of Being Prey'. *Quadrant*. 29(3): 29–34.
Plumwood, Val (2007a). 'The Cemetery Wars: Cemeteries, Biodiversity and the Sacred'. *Local-Global: Identity, Security, Community*. 3: 54–71.
Plumwood, Val (2007b). Review of Deborah Bird Rose's 'Reports from a Wild Country: Ethics for Decolonisation'. *Australian Humanities Review*. 42. www.australianhumanitiesreview.org/2007/08/01/a-review-of-deborah-bird-roses-reports-from-a-wild-country-ethics-for-decolonisation, accessed 22 November 2023.
Plumwood, Val (2012). *The Eye of the Crocodile*, ed. Lorraine Shannon. Canberra: ANU Press.
Spedalieri, Francesca (2014). 'Quietly Posthuman: Oriza Hirata's Robot-Theatre'. *Performance Research*. 19(2): 138–140.
van Baarle, Kristof (2019). 'Void'. In Peter Eckersall and Helena Grehan, eds, *The Routledge Companion to Theatre and Politics*. London: Routledge, pp. 228–231.
van Baarle, Kristof (2020). 'End-Time Attitudes: Performing the Last Part'. In Kristof van Baarle and Peter Eckersall, eds, *Machine Made Silence: The Art of Kris Verdonck*. Aberystwyth, UK: Performance Research Books, pp. 152–162.
Verdonck, Kris (2016), *IN VOID*. Performance Work, A Two Dogs Company.
Verdonck, Kris (2017). *CONVERSATIONS (at the end of the world)*. Performance Work, A Two Dogs Company.
Verdonck, Kris (2019). *SOMETHING (out of nothing)*. Performance Work, A Two Dogs Company.
Verdonck, Kris (2023). *PREY*. Performance Work, A Two Dogs Company.

4 Ear

We begin with vibration. The back and forth of minute particles in the air. This vibration is a pattern of disturbance that creates pressure, a change in atmosphere that becomes a wave. When this wave hits the outer edges of the human ear, the pinna, it is collected and filtered, along with other waves, and channelled into the ear canal. Next, the waves hit the tympanic membrane, the eardrum, which vibrates anew. The wave is amplified and sets three of the smallest bones in the body into motion. The motion of the bones, the operation of the drum, amplifies the energy of the wave as it enters the inner ear, the cochlea. The cochlea is shaped like a snail. Inside the snail shape is a spiral. Inside the spiral are tiny hair cells that respond to different frequencies. As the pressure waves move through the inner ear, hair cells and stereocilia transform the vibrations into electrical energy (Kelly et al. 2005; Sundar et al. 2021). A change in atmosphere creates vibrations that result in pressure that becomes energy that then dissipates, steadying the air until the arrival of new vibrations. Hearing is the performance of bodies and brains making sense of vibrations moving through other bodies of air, water, earth, rock, flesh.

To consider vibration and its sociocultural significance, this chapter focuses on examples of biological and technological processes of cellular interaction and disintegration in contemporary performance. Specifically, we examine the ear and different technologies of 'hearing', 'listening' and what Rebecca Collins describes as 'aural attunement', which attends to the frequencies and rhythms of the more-than-human using the whole body (rather than the isolated organ of the ear) in order to 'orientate our relation to the world' (2021). We offer this analysis through the works of Cypriot-Australian artist Stelarc's *Extra Ear* project (1997–present) and collaborations between Deaf and hearing artists in Australian dance-theatre company The Delta Project's *Under My Skin* (2016). We focus on the movements of air, breath, growth and decay that are embedded in the acoustics of the inner ear and the technologies that intervene in the reception of vibrations, from cochlear implants to zoom lags, to hearing loops and signed performance. From there, we explore Indigenous philosophies and practices of listening to and being part of Country in Australian Indigenous culture. We reflect on the invitation to learn about Indigenous Australian song-spirals as a particular mode of listening to and with Country from the Gay'wu

DOI: 10.4324/9781003273080-4

Group of Women from North East Arnhem Land (2019). Drawing on recent technological developments and ancient knowledge shared from First Nations cultures, we elaborate on performing climates as a dramaturgy of atmospheres, actors and affects that reconstitute relations between biological, technological and climatic conditions across the stage (Paterson 2019).

The Extra Ear

Ear on Arm is part of Stelarc's iterative *Extra Ear* project which began in 1997 and continues today. To create *Ear on Arm*, the artist collaborated with a team of medical surgeons to construct an extra ear on his left arm (Stelarc 2023a). *Ear on Arm* is a soft prosthesis constructed from a 'Medpor implant', a porous, biocompatible polyethylene material over which excess skin was grown and stretched to make the ear shape (Stelarc in Zylinska 2002: 125). Once the 'scaffold' was surgically inserted, the 'ear' slowly knitted with Stelarc's cells to establish its own blood supply. The work was partly inspired by the Vacanti mouse, a well-known science experiment released in 1997 where a lab mouse had what looked like a human ear grown onto its back. The Vacanti research aimed to experiment with new developments in tissue engineering in order to grow human body parts for grafting, but the photo of the mouse in the popular media caused international alarm about the implications of genetic engineering (Kruszelnicki 2006). Stelarc, who has long been interested in the relationship between bioengineering, experimentation, medicine, technology and performance art, has frequently used his own body to experiment with human architecture (Figure 4.1).

We can see here an image of the ear shape on Stelarc's left arm, taken in 2020, with its distinctive folds and curves. The image of the arm shows the ear skin crisscrossed with tiny wrinkles. These wrinkles show that the ear on the arm is distinct from the smooth and tight skin that might typically cover a human ear. There is no visible cavity for an ear canal and Stelarc's extra ear also lacks a middle and inner ear. There is no eardrum for the air to pass through and no stereocilia and cochlear to respond to pressure waves and modify the frequency of vibrations in the air. In short, the sounds that pass across Stelarc's extra ear are limited to interaction with skin cells and the more robust hair cells found on a human arm. This ear is not a working replica.

As Pratap Sundar et al. note: 'The human ear has been a model for designing many audio devices, e.g., microphone design, middle ear prosthetics, hearing-aid design, cochlear implants, miniature hair bundle sensors, robot ear design [and] robot locomotion' (2021: 12). Significantly, however, in *Ear on Arm*, Stelarc explores the sculptural possibilities of tissue engineering without the kind of utilitarian goals of engineers, audio designers or creators of the Vacanti mouse. Stelarc evokes the human ear but deliberately overlays the aesthetic form with different functions. He writes:

> Imagine an ear that cannot hear but rather can emit noises. Implanted with a sound chip and a proximity sensor, the ear would speak to

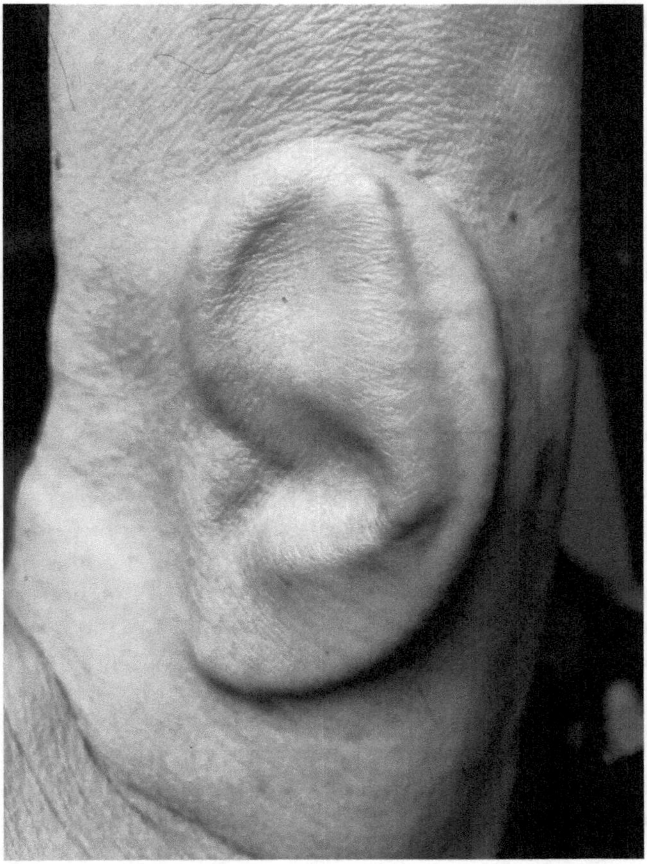

Figure 4.1 Stelarc: *Ear on Arm* (detail), 2020.
Photo: Nina Sellars.

> anyone who would get close to it. Perhaps, the ultimate aim would be for the Extra Ear to whisper sweet nothings to the other ear.
> (Stelarc 2023b)

Stelarc emphasises that his excess ear performs a 'function' that is contradictory to its normative functionality: he desires the creation of an ear that not only does not listen but also one that speaks. Indeed, *Ear on Arm* was designed to involve the integration of microphone and Bluetooth technology into the 'ear' so that it could capture and transmit sound:

> An extra ear is presently being constructed on my arm. A left ear on a left arm. An ear that does not hear but transmits. A facial feature had been replicated, relocated and repositioned elsewhere ... When electronically augmented the ear will become internet enabled in any wifi

hotspot, becoming a mobile, accessible remote listening device for people in other places.

(Stelarc 2019: 13)

It is striking that this project began in the late 1990s, prior to the ubiquity of smart phones/watches, the Internet, and before the intensity of surveillance culture that took hold post the New York terror attacks of 11 September 2001. Given the hyperconnected nature of the world today and the invasiveness of what Shoshana Zuboff (2019) terms 'surveillance capitalism', the act of listening in to a remote site through Stelarc's *Extra Ear* invites a more sinister reading than what was suggested by its conceptual model in 1997. On the one hand, we know, particularly since the COVID-19 pandemic, that communication technologies such as Zoom, FaceTime and Microsoft Teams can provide us with access to audio-visual connection in ways never before conceived. These platforms have been particularly advantageous for people with disabilities, lack of financial means or primary caregivers as well as people who juggle intersectional demands such as Women of Colour, First Nations people and people living in the Global South. On the other hand, beyond everyday zoom fatigue, lagging audio and the risks of isolation, Zuboff notes that the stripping away of privacy protections, mass data scraping and the oversaturation of information in the long-term exacerbates income disparity and deepens technological and political divides (2019).

In 2007, Stelarc addressed questions of information overload and human alienation in ways that are even more strikingly relevant today, noting that it is a qualitative as much as quantitative problem for humans in the contemporary world. He states: 'information overload is a situation where the technology expands the electromagnetic sensing of the body and hence the body has to constantly generate paradigm shifts to understand itself in the world' (Abrahamsson and Abrahamsson 2007: 295). Stelarc further reminds us that the human ear is not only an organ of hearing but also one that is essential for our bodies to maintain balance. Therefore, an extra ear suggests a re-orientation of the body (Stelarc 2005: n.p.). *Ear on Arm* is one way of the body experimenting with the re-negotiation of itself in the world through its provocative reimagining of hearing, listening, speaking within particular geographies and spatialities of the body and its environment within linear, Western notions of time.

Stelarc makes clear that the bioengineered extra ear does not signify a lack or deficiency of the human body, but rather an excess. Long-time collaborators and bioartists Oron Catts and Ionat Zurr note that:

> Rather than replacing a missing or malfunctioning part of the body, these artefacts [detached bioengineered 'ears'] are alternate additions to the body's form and function. Stelarc refers to his body in performance as The Body, the obsolete human body penetrated and modified by technology.
>
> (2008: n.p.)

Disability theorists have long been grappling with the continuing perversions of the medical model of disability in which 'malfunctioning' parts are to be 'fixed' and modified by technology. Indeed, Margrit Shildrick points out the heteronormative history of prosthetic practice and the risks of the 'trope of enhancement' inherent in claims for posthuman 'progression' (2017: 144). Yet, as Julie Clarke recognises, rather than presenting posthuman perfection, the *Extra Ear* project also uses an aesthetics of the abject, one which invokes not only fear of the other but also fear of difference or mutation, offering a potential commentary on the aftereffects of pollution, radiation, chemical poisonings and other environmental damages caused by humans (Zylinska 2002: 37). This engagement in the abject also recalls the popular horrified response to images of the Vacanti mouse as a 'Frankenstein' experiment. Yet *Ear on Arm* is not a prosthetic enhancement that aims for normative perfection but a way of questioning how the body functions and looks. The alienated and unsettled ear, a new acoustic vehicle with different abilities, recasts the human ear as one technology among a multitude of different types of hearing, speaking and listening in performance.[1] In order to explore further multimodal dimensions of vibration and the ear, we now turn to the innovations of d/Deaf dance theatre.

The Ear That Hears Differently

Stelarc's *Ear on Arm* is only a representation of a human ear, an ear that does not function as human ears in able bodies are 'supposed' to function. The title of Melbourne-based ensemble The Delta Project's *Under My Skin* (2016) also draws attention to what is hidden under the skin of the human body, to make the question of how humans process sound far more complex through performance with nontraditionally hearing human bodies. *Under My Skin* is a collaboration between Deaf and hearing artists in The Delta Project that offers a glimpse into how performance might reimagine listening and hearing across an expanded range of communicative modes. Co-founded in 2012 by Deaf dance artists Jo Dunbar and Anna Seymour, The Delta Project's work explores the merging of Deaf and hearing artists and their experiences. *Under My Skin* began with an Auslan-led choreographic approach in which Auslan, the majority sign language of the Australian Deaf community, became a primary means of communication used by ensemble members.

In the performance *Under My Skin*, Auslan remained a key communicative mode, but the signed language also intermingled with the visual and kinaesthetic characteristics of dance theatre. Dunbar (d/Deaf) and Lina Limosani (hearing) choreographed the work, which opened on the image of entangled human bodies set against a background of photographic portraits of the dancers' faces. The bodies twist and writhe as the images, designed by new media artist Rhian Hinkley, pan across them. As they merge with the bodies, the images distort and fragment. Slowly, individual dancer bodies emerge, and the four ensemble members become more physically distinct from one another and the shifting background.

In order to ensure synchronicity throughout this performance, the dancers relied heavily on visual cues, face-to-face interactions and a sound design from Russell Goldsmith that used lower frequencies, a seat-shaking score with lots of rumbling bass. As such, The Delta Project's approach foregrounded Deaf signing within, as Gabrielle Hodge describes, a complex visual, kinaesthetic, symbolic and situated 'language ecology' (2020: 62). Indeed, this multimodal ecology is also a fundamental part of d/Deaf experience, and thus a unique innovation highlighted in The Delta Project's decision not only to centre Auslan as a means of communication for the ensemble but to have moments when the dancers signed to each other and to the audience.[2] For Hodge:

> Using a signed language, or learning to interact using a signed language because you are working with people who do, means adapting to the differences presented by deaf bodies — it means growing yourself into the collective 'skin' of deaf people who sign.
>
> (2020: 78)

The importance of skin, the skin on Stelarc's arm and the metaphorical title of The Delta Project's performance, draws attention to the ways in which atmosphere and vibration interact with not only the human ear but across all sorts of human and more-than-human skin surfaces.

The affective and ecocritical connections to the sensory experience of theatre are the subject of much scholarship (Welton 2012; Lavery 2019), including two recent edited volumes on *Staging Atmospheres* (Déchery and Welton 2020, 2021). Included in the second volume is Rebecca Collins' analysis of Emma Bennett's *Slideshow Birdshow* (2013), a work that explores human and avian relations. Here, Collins writes of 'Affective Attunement' as a means of listening with the whole body that encompasses 'a sensorial mode of engagement with the world, one that lies outside classifying structures and systems of representation' (2021). Collins cites philosopher Michel Serres, who writes:

> [W]e hear through our skin and feet. We hear through our skull, abdomen and thorax. We hear through our muscles, nerves and tendons. Our body-box, strung tight, is covered head to toe with a tympanum.
>
> (2016: 141)

The vibratory, physiological and atmospheric qualities of hearing described by Serres were readily experienced in the sound design for *Under My Skin*. As mentioned, Goldsmith utilised low frequencies, and this score was created for both hearing and nonhearing dancers and spectators. Working within the range of 25 and 100 hertz meant that Goldsmith's score would contain vibrations that would be 'felt' by the human body, in organs and across the skin, rather than relying on the eardrum, to engage a different perceptual range (Ross 2016).

While the use of signed language in *Under My Skin* was a primary ingredient of the choreographic approach, vibration according to wavelength, amplitude, frequency, time period and velocity across parts of the body not traditionally associated with 'hearing' link Stelarc's electronic extra ear to The Delta Project's visual and kinaesthetic listening. What these performances show is the porousness of human experiences of hearing and feeling that the ear is not all that 'hears' in the human body. Reflecting on the experience of watching *Under My Skin*, two elements remain manifest to us as nonsigning, hearing spectators. As described earlier, these elements are the use of the body as a 'hearing' technology and being immersed in a complex language ecology. There is a material and emotional aspect to this, feeling is responsive. In the words of Hodge:

> This involves developing one's sensitivity to the visual and kinaesthetic dynamics of an interaction (sign, dance, light, shadow), such as through attentive 'listening' (including meaningful use of eye gaze), and expanding one's communicative repertoire and awareness of different semiotic processes. These qualities may be revealed to non-signers in different ways, across different time depths.
>
> (2020: 78)

The Delta Project's merging of Deaf and hearing worlds reveals the use of an extended range of affective attunement that uses the entire body to hear.

Considering the human body as an acoustic technology and taking into account the various ways in which humans and more-than-humans may 'hear', it is also incorrect to describe, as Stelarc does above, the extra ear as one that only transmits and 'does not hear'. Instead, we would suggest that Stelarc's ear in fact 'hears differently'. This is indicative of him working with normative understandings of what it means to 'hear', linked to a Cartesian idea of a mechanical body as split from emotional processing. Indeed, writing in response to Shildrick's work on prosthetics, disability and transcorporeality, Moritz Ingwersen considers:

> transgressive corporeality not as a marker of pathology, but as a mode of being that can be instructive in reconsidering the customary perception of the relationship between embodied agents and the world.
>
> (2017: 167)

Ingwersen suggests that perceptual relations between human and nonhuman may be extended by new understandings of affective relations that emerge 'when human agents recognize their energetic openness instead of idealizing closure' (2017: 167). In this way, a vibratory change that alters embodied experience also opens ways of rethinking interactions between human and more-than-human. To reflect on 'new' ecologies of hearing and listening we turn to ancient and contemporary practices of our First Nations people. We now take up an invitation to explore Indigenous Australian ways of knowing and the experience of listening to Country to learn how these perceptual, vibratory and climactic entanglements might be extended across all forms of life.

The Ear That Listens to Country

The act of listening to one another or to an 'other' that we've described so far are mechanical, technological or biological processes. But the act of listening is also a performance of respect in human culture, part of an ethics of attentiveness to the other and a sign of power relations – those to whom we listen are those whom we respect. To be heard is to be respected, to be acknowledged and valued. As settler scholars, trained in a Western philosophical tradition and overwhelmingly Western traditions of theatre and performance, the act of listening to the more-than-human world is rare and outside the knowledge systems with which we are most familiar. Notably, the artificial split between the 'two cultures' of science and arts have erased potentially useful opportunities for cross-pollination of knowledge and even a means to 'listen to' or be attentive to the more-than-human.

This kind of oversight and ignorance has come hand-in-hand with ignorance of Indigenous knowledge which has, at best, been given attention as a binary critique of Western philosophies (see, e.g. Langton 1995; Moreton-Robinson 2000, 2013; Watson 2014), and, at worst, derided as nonscientific and nonrational through erasure or exclusion. In an attempt to both overturn this racist history of philosophical thought and also to expand a critical understanding of listening, we give our ears, metaphorically, and our critical engagement, to Indigenous voices in a gesture of respect, curiosity and openness to vital and urgent learnings from Indigenous knowledge. As Indigenous-Australian scholar and writer Ambelin Kwaymullina (Palyku) writes:

> My suggestion is that non-Indigenous scholars must respect Indigenous sovereignty and meaningfully enact this respect, including through the layered process of listening to the voices of Indigenous women.
>
> (2018: 193)

In particular, we take up the invitation to learn about Indigenous Australian songspirals from the Gay'wu Group of Women from North East Arnhem Land, one of the most northern tips of Australia. These women note that although the knowledge they share is their property, they give permission for us to write about and reflect upon it but not be authorities on it (2019: xxv). In doing so, they ask that we honour the context of the songspirals and the layers of knowledge, noting that some knowledge is not for sharing outside their clans and that our comprehension will only ever be superficial, surface, theoretical rather than a lived understanding of it (2019: xxv–xxvi).

In explaining songspirals, the Gay'wu Women begin by clarifying the concept of 'Country'. They write:

> Country gives the knowledge for this book. It guides us and teaches us. Country has awareness, it is not just a backdrop. It knows and is part of us. Country is our homeland. It is home and land, but it is more than that. It is the seas and the waters, the rocks and the soils, the animals and winds and people too. It is the connections between those beings, and their dreams and emotions, their languages and their Law. Country is the way humans and non-humans co-become, the way we emerge together, have always emerged together and will always emerge together.
>
> (2019: xxii)

Such a description of Country immediately reveals a radically different understanding of the Western backgrounding of 'environment', to a living, breathing entity that has agency and even potentially something approximating consciousness.

From there the Gay'wu Women define the practice of a songspiral, something that is co-created within Country that is alive. They write:

> Every songspiral is a song, a ceremony, a picture, a story, a person, a place, a mapping, some things that we did, that we do and that we will always do, and it is all of those things together because those things are really the same. And it is more, much more. It is knowledge and language and Law.
>
> (2019: xxv)

The songspiral is thus not a song with music and/or lyrics, as in a Western understanding, but a weaving of everything together – knowledge, ritual, maps, a guide, family, past and present and Law – it is something which connects kin and homeland to waters, rocks, winds and everything else (2019: xxii). As outsiders to this knowledge, it is interesting that this weaving of the human and more-than-human occurs through the evocation or translation of the idea of song – music, sound or vibration as well as a shape – the spiral, that mirrors the formation of the ear as a kind of coiled organ, drawing vibrations downwards.

The notion of Listening to Country is not necessarily accessible to settlers, yet it seems clear that it extends the concepts of 'hearing' and 'listening' beyond the organ of the human ear, beyond the perceptual range of the human body, into vibrations that interact throughout human and more-than-human worlds across the past-present-future times of Indigenous Australian experience and culture. Attempting to apply our learnings drawn from the Gay'wu Group of Women, the final section of this chapter situates performing climates as a critical response to the colonial frameworks that have dominated relations for the last few hundred years. We now turn to the performances of the wood ear fungus in order to extend the perceptual range of the ear and its relationships between human and more-than-human performance.

Country That Hears

The wood ear fungus or *Auricularia auricula-judae* is a gelatinous European fungus with an elastic texture when fresh (Figure 4.2). It grows on old or dead wood of deciduous trees (particularly Elder trees) in a form resembling a human ear. The underside of *A. auricula-judae* is sometimes folded or wrinkled and has coloured line markings that resemble veins, making it appear even more ear-like. The name *A. auricula-judae* comes from the story of Judas Iscariot hanging himself from an Elder tree and the fungus is commonly described as 'Judas's ear' or 'Jew's ear', a corruption that stems from the idea that Judas was Jewish. The Japanese call it 'tree jellyfish' and the Chinese, who prize it both for its use in cuisine and in Chinese medicine, call it 'wood ear' for its distinctive ear and jelly shape (Luo et al. 2009). The variety of names suggest the evocative, creative responses that the aesthetics of the fungus inspires, the intertwining of histories and historical figures, mythologies, animal species and organic human shapes.

The wood ear fungus is high in nutrients and is commonly used in Chinese cuisine in salads and stir fries. As such, today it has a large commercial production, particularly in rural China and is valued for its composting capacities, converting agricultural waste to food and creating jobs in the provinces

Figure 4.2 Wood ear fungus.

in China (Zhang et al. 2012). It was used in medicine prior to the nineteenth century and more recently, has been proven to have wound-healing properties (Mapoung et al. 2021) and antimicrobial activities against *Escherichia coli* and *Staphylococcus aureus* (Cai et al. 2015). Its pharmacological properties show promise for the potential development of anti-microbial proteins that are valuable to humans given the growing resistance of bacteria to antibiotics (Oli et al. 2020). It is thus, as Anna Lowenhaupt Tsing writes of the Matsutake fungus (2015), deeply embedded in the particular cultural, biological and ecological lives of humans through its culinary value, as well as medicinal treatments for bodily health and survival, that have been disrupted by human mishandling.

The aesthetics of the wood ear fungus can be viewed anthropocentrically as resembling a human ear, but like Stelarc's *Ear on Arm* or the lived-experiences of the Deaf performers from The Delta Project, it is an ear that does not function in normative ways, and thus it unsettles the human attributions of things and their functioning. Instead, it engages in its own complex performance within ecological systems of growth, composting and decay that humans are only beginning to understand. Like many fungi on the planet, the wood ear serves a vital function as a decomposer, a quiet bacterial symphony that happens via its root systems which facilitate the breakdown of matter on the forest floor, performed without care or concern for humans (Lowenfels 2017).

Performing climates invites us to see the cellular interplay between fungi, bacteria, soil and root as having a dramaturgy of its own devising. This is a performance that is not for humans. Instead, it uses a slow dramaturgy that does not exist at the Western human pace and sits well outside the realm of 'anthropocentric performance' (Stevens 2019a). Nevertheless, Indigenous Australian approaches to Listening to Country suggest that recognition and constant care of more-than-human beings does not need to be disconnected from human understanding. Within the realm of performing climates, the wood ear fungus holds its own knowledge, language, communication systems and Law. In keeping with the function of most human ears (hearing and nonhearing), viewing the dramaturgy of the wood ear fungus 'as' performance, rather than a species to be used for human consumption or medicine, momentarily throws us off-balance, destabilises our habituated anthropocentric perspective. Perceiving the wood ear fungus as a performance of more-than-human vibrations (hearing, listening, speaking) can potentially reorient our position in the biosphere in which we are one tiny actor among many. Human life, human theatre and human hearing is only possible through more-than-human vibrations, growth, disintegration and decay.

In focusing on human and more-than-human elements in Stelarc's *Extra Ear* project, The Delta Project's *Under My Skin*, Indigenous Australian songspirals and the wood ear fungus, invites new configurations of biological and technological processes of cellular growth and vibration. These performing climates allow us to expand the capacity of performance to include a dramaturgy of more-than-human atmospheres, actors and affects.

Notes

1 Here we recall Mel Brooks' parody of Mark Antony's speech from Shakespeare's *Julius Caesar*: 'Friends, Romans, countrymen, lend me your ears'. In Brooks' version, in the 1993 film *Robin Hood: Men in Tights*, the listeners rip off their prosthetic ears and throw them at him. The symbol of the ear has become a metaphor for listening, but Brooks literalises the act of borrowing a detachable ear in a parody of power and status that acts of listening assume.

2 Deaf scholar and theatre artist Racheal Missingham traces the rich cultural legacy of Australian d/Deaf theatre across 50 years and notes the vital role of Deaf-led practices and companies as a way of exploring d/Deaf experience, as well as collaborations with hearing-allies, to support the unique cultural production of d/Deaf artists (2021).

References

Abrahamsson, Christian and Sebastian Abrahamsson (2007). 'Cultural Geographies in Practice: In Conversation with the Body Conveniently Known as Stelarc'. *Cultural Geographies*. 14(2): 293–308.

Cai, Ming, Yang Lin, Yin-long Luo et al. (2015). 'Extraction, Antimicrobial, and Antioxidant Activities of Crude Polysaccharides from the Wood Ear Medicinal Mushroom *Auricularia auricula-judae* (Higher Basidiomycetes)'. *International Journal of Medicinal Mushrooms* 17(6): 591–600.

Catts, Oron and Ionat Zurr (2008). 'The Art of the Semi-Living and Partial Life: Extra Ear – 1/4 Scale'. In Melentie Pandilovski, ed, *Art in the Biotech Era*. Adelaide: Experimental Art Foundation, pp. 140–147.

Collins, Rebecca (2021). 'Aural Attunement Is a Form of Performative Writing Used to Write About Nonhuman Affective Relations in Emma Bennett's *Slideshow Birdshow* (2013)'. *Ambiances*. 2: 1–20.

Déchery, Chloé and Martin Welton, eds (2020). 'Staging Atmospheres: Theatre and the Atmospheric Turn'. *Ambiances* 1: Special Issue.

Déchery, Chloé and Martin Welton, eds (2021). 'Staging Atmospheres: Theatre and the Atmospheric Turn'. *Ambiances* 2: Special Issue.

Gay'Wu Group of Women (2019). *Songspirals: Sharing Women's Wisdom of Country through Songlines*. Sydney: Allen & Unwin.

Haraway, Donna (1991). *Simians, Cyborgs, and Women: The Reinvention of Nature*. New York: Routledge.

Hodge, Gabrielle (2020). 'The Ideology of Communication Practices Embedded in an Australian Deaf/Hearing Dance Collaboration'. In Annelies Kusters, Mara Green, Erin Moriarty and Kristin Snoddon, eds, *Sign Language Ideologies in Practice*. Berlin: De Gruyter Mouton, pp. 59–82.

Ingwersen, Moritz (2017). '*Cybernethics*: Thinking Bodies and Boundaries through Science'. In Anne Waldschmidt, Hanjo Berressem and Moritz Ingwersen, eds, *Culture – Theory – Disability: Encounters between Disability Studies and Cultural Studies*. Bielefeld, Germany: Transcript, pp. 160–169.

Kelly, Matthew W., Doris K. Wu, Arthur N. Popper and Richard R. Fay, eds (2005). *Development of the Inner Ear*. New York: Springer.

Kruszelnicki, Karl (2006). 'Mouse with Human Ear'. *ABC Science,* Australian Broadcasting Corporation, 2 June. www.abc.net.au/science/articles/2006/06/02/1644154.htm, accessed 5 July 2023.

Kwaymullina, Ambelin (2018). 'You Are on Indigenous Land: Ecofeminism, Indigenous Peoples and Land Justice'. In Lara Stevens, Peta Tait and Denise Varney, eds, *Feminist Ecologies: Changing Environments in the Anthropocene*. London: Palgrave Macmillan, pp. 193–208.

Langton, Marcia (1995). 'The European Construction of Wilderness'. *Wilderness News*, Summer 1995–1996: 16–17.

Lavery, Carl, ed (2019). *Performance and Ecology: What Can Theatre Do?* London: Routledge.

Liboiron, Max (2021). *Pollution Is Colonialism*. Durham, NC: Duke University Press.

Lowenfels, Jeff (2017). *Teaming with Fungi*. Portland, OR: Timber Press.

Luo, Yangchao, Gang Chen, Bo Li, et al. (2009). 'Evaluation of Antioxidative and Hypolipidemic Properties of a Novel Functional Diet Formulation of *Auricularia auricula* and Hawthorn'. *Innovative Food Science and Emerging Technologies*. 10(2): 215–221.

Mapoung, Sariya, Sonthaya Umsumarng, Warathit Semmarath et al. (2021). 'Skin Wound-Healing Potential of Polysaccharides from Medicinal Mushroom *Auricularia auricula-judae* (Bull)'. *Journal of Fungi*. 7: 247.

Missingham, Racheal (2021). 'Australia's Deaf Theatre: Past, Present and Future'. Master of Philosophy Thesis, Queensland University of Technology.

Moreton-Robinson, Aileen (2000). *Talkin' up to the White Woman: Aboriginal Women and Feminism*. St. Lucia: University of Queensland Press.

Moreton-Robinson, Aileen (2013). 'Towards an Australian Indigenous Women's Standpoint Theory: A Methodological Tool'. *Australian Feminist Studies*. 28(78): 331–347.

Neale, Margaret (2022). 'First Knowledges: An Introduction'. In Zena Cumpston, Michael-Shawn Fletcher and Lesley Head, eds, *Plants: Past, Present and Future*. Melbourne: Thames and Hudson, pp. 1–4.

Oli, Angus Nnamdi, Peter Anyigor Edeh, Reham M. Al-Mosawi et al. (2020). 'Evaluation of the Phytoconstituents of *Auricularia auricula-judae* Mushroom and Antimicrobial Activity of Its Protein Extract'. *European Journal of Integrative Medicine*. 38: 101176.

Paterson, Eddie (2019). 'Compost and Air-Conditioning: Beyond Biospherical Performance and towards the Shimmer'. *Performance Research*. 24(8): 31–36.

Ross, Annabel (2016). 'Deaf Can Dance: The Delta Project Combines Hearing and Non-Hearing Dancers in Under My Skin'. *Sydney Morning Herald*, 29 April. www.smh.com.au/entertainment/dance/deaf-can-dance-the-delta-project-combines-hearing-and-nonhearing-dancers-in-under-my-skin-20160429-goinls.html, accessed 1 November 2023.

Serres, Michel (2016). *The Five Senses: A Philosophy of Mingled Bodies*. London: Bloomsbury.

Shildrick, Margrit (2017). 'Border Crossings: The Technologies of Disability and Desire'. In Anne Waldschmidt, Hanjo Berressem and Moritz Ingwersen, eds, *Culture – Theory – Disability: Encounters between Disability Studies and Cultural Studies*. Bielefeld, Germany: Transcript, pp. 137–151.

Stelarc (2005). 'From Zombie to Cyborg Bodies – Extra Ear, Exoskeleton and Avatars'. *NeMe*. www.neme.org/texts/from-zombie-to-cyborg, accessed 22 November 2023.

Stelarc (2019). 'Extra Ear: Alternate Anatomical Architectures'. stelarc.org/_php, accessed 11 December 2023.

Stelarc (2023a). 'Ear on Arm'. http://stelarc.org/_activity-20242.php, accessed 11 December 2023.

Stelarc (2023b). 'Extra Ear'. stelarc.org/_activity-20229.php, accessed 11 December 2023.

Stevens, Lara (2019a). 'Anthroposcenic Performance and the Need for "Deep Dramaturgy"'. *Performance Research*. 24(8): 89–97.

Sundar, Pratap, Chandan Chowdhury and Sagar Kamarthi (2021). 'Evaluation of Human Ear Anatomy and Functionality by Axiomatic Design'. *Biomimetics*. 6(2): 31.

Tsing, Anna Lowenhaupt (2015). *The Mushroom at the End of the World*. Princeton, NJ: Princeton University Press.

Under My Skin (2016). Performance Work, The Delta Project.
Watson, Irene (2014). 'Re-Centring First Nations Knowledge and Places in a Terra Nullius Space'. *AlterNative: An International Journal of Indigenous Peoples*. 10(5): 508–520.
Welton, Martin (2012). *Feeling Theatre*. Basingstoke: Palgrave Macmillan.
Zhang, Rui Ying, Dan Dan Hu, Jin Gang Gu et al. (2012). 'Development of SSR Markers for Typing Cultivars in the Mushroom *Auricularia auricula-judae*'. *Mycological Progress* 11: 587–592.
Zuboff, Shoshana (2019). *The Age of Surveillance Capitalism*. London: Profile Books.
Zylinska, Joanna, ed (2002). *The Cyborg Experiments: The Extensions of the Body in the Media Age*. London: Continuum.

5 Ants

> Today, the décor, the wings, the background, the whole building have come on stage and are competing with the actors for the principal role. This changes all the scripts, suggests other endings. Humans are no longer the only actors, even though they still see themselves entrusted with a role that is much too important for them.
>
> (Latour 2018: 43)

We begin with ants. Or rather, we begin with a single worker ant, one belonging to the *Atta* genus. Commonly known as a leafcutter ant, it carries organic plant material along a subterranean route towards the centre of its nest. The organic material is freshly cut. It consists mainly of leaves that will be fed to the fungi gardens of the ant society. The gardens are deep below the nest and the fungi, all of which are members of the *Lepiotaceae* family, are actively cultivated by the ants that feed them and keep their gardens free of pests and mould. The fungi were domesticated 15 million years ago, after a process that lasted 30 million years. The ants and the fungi have another symbiotic partner here, a bacterium that grows on the ants and aids their fungus-based agriculture by introducing antimicrobials into the system (Figure 5.1).

The ant delivers the leaf matter to the garden workers (that are its younger, healthier counterparts) and then turns to gather a load of waste. This waste is then transported to the peripheries of the nest, to one of many waste dumps in which refuse is organised carefully into chambers. Once dropped at the chamber, heap workers aid the process of decomposition by turning and reorganising the matter. In the process of breakdown and decay – the composting – carbon dioxide is produced. Due to the heat of the gas, and the orientation and design of the chambers, vapour flows between the different chambers with different atmospheric pressures and velocities of air. The warm carbon dioxide is expelled through earth chimneys above, just as fresh air is drawn down into the underground megalopolis. The superorganism of the ant colony relies on this careful arrangement of waste, fungi, bacteria and air. It might be said that the ants create an advanced compost air-conditioner from natural materials.

DOI: 10.4324/9781003273080-5

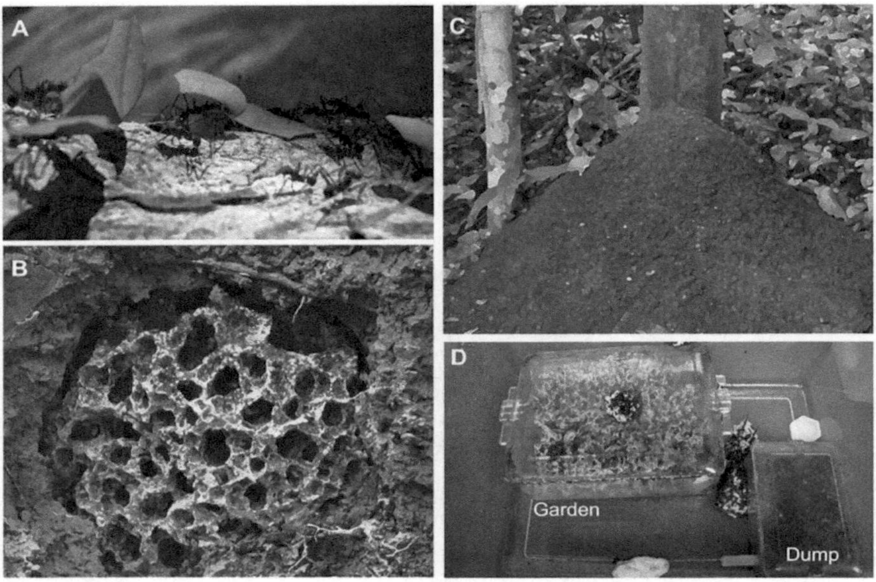

Figure 5.1 Leafcutter colony composting system. (A) Ants gathering leaves. (B) Fungus and plant material in colony. (C) Pile of discarded fungus. (D) Typical lab colony set-up showing fungus garden and refuse dump.

Images courtesy of Creative Commons.

Western theatre and performance today is still predominantly located in biospheres, indoor theatre spaces, where cycles of growth and decay are controlled by air-conditioning or heating, or rendered invisible or insignificant. Yet, under our feet, in the ant's nest, the dramaturgy of biomass plays out.

'Another heavenly day' begins Winnie, in Samuel Beckett's 1961 play *Happy Days*. Yet, Winnie's line is not the start of the performance. *Happy Days* begins with a pile of soil:

> Expanse of scorched grass rising centre to low mound. Gentle slopes down to front and either side of the stage. Back an abrupter fall to stage level. Maximum of simplicity and symmetry.
>
> Blazing light.
>
> (Beckett 1990: 138)

Consider the mound not as a piece of scorched earth, or denuded grass, but as a dynamic living system – capable of changing its material composition according to patterns and rates of erosion – a catalyst of plant nutrition, a terrestrial ecosystem in which billions of bugs and microorganisms can live within a single handful of 'dirt'. As David R. Montgomery writes, soil is 'the skin of the earth' and the histories of human civilisations have depended on the health and amount of soil available to them (2008: 13–16). Maybe we've been concentrating on the play's central human figure of Winnie for too long – perhaps it's time to rethink *Happy Days*, time to imagine what will happen to Winnie as she stays in the dirt, and as her body begins to form relations with bacteria and acts as fertiliser, feeding the soil and its teeming biomass.

French director Claire Denis' sci-fi film *High Life* (2018) also begins with soil. A garden. Lush plants, moist leaves and a lingering pan over a vegetable plot. Then, the whirring noise of an air-conditioning system on a spaceship, on which a group of criminals are trapped, with the mission to extract energy from a black hole. Yet, in contrast to *Happy Days* and to earlier versions of intergalactic horticulture, *High Life* moves away from the biospherical mode. It rejects the human in favour of the bacterial and organic. It is through the character of Tcherny (André Benjamin), a prisoner who becomes increasingly obsessed with the garden and takes to walking on the soil with his bare feet. Around three-quarters of the way into the film, Tcherny takes his own life in the garden and lies down on the soil. Here, the camera lingers again as the human body is absorbed into the earth. As it fades, the body seems to shimmer.[1] The vegetables continue to grow, and we see them harvested long after Tcherny has become compost.[2]

Insect Agency

In *Termitaria* (2001), contemporary environmental artist Perdita Phillips collaborated with termites. For this artwork, Phillips buried 850 romance novels in soil for a year. The insects, commonly known as white ants that consume decaying materials buried underground, ate the books.[3] In so doing, the insects were also likely intermingling with nematodes, moulds and other microscopic organisms in the soil. While many of the books were completely consumed, Phillips was able to 'excavate' some remaining works and display them as 'sculptures'. These half-eaten copies look as though they might have been buried centuries ago, the edges of the pages undulate with holes and gaps, and only the spines of the books retain their structure.

However, while their forms are indeed sculptural due to the layers of pages, arrangements of the holes and aging patina, *Termitaria* is less a work for humans than a multiscalar engagement. The novels are now part of a 'romance' with the soil, a more-than-human sharing of cells. Notably, in a later work, *Termite Embassy* (2015), Phillips playfully engages with notions of termite agency and rights by placing tiny flags into termite mounds. Taking place in the lead-up to the COP21 negotiations in Paris in 2021, and recalling Laurence Tubiana, Bruno Latour and Frédérique Aït Touati's performance event *Climate – Make It Work* (2015), this work asks how the termites might intervene or participate in global diplomatic summits in response to human-caused climate change.

In Winnie's mound, Tcherny's garden, the leaf cutter and termite mounds, the microbial energy of the soil continues because there is organic matter and creatures to interact with. Indigenous Australians have long nurtured complex engagement with termites. On Gurindji Country, in Northern Australia, termites are prized for their use in medicine – particularly to aid the health of newborn babies as well as for environmental management and as a sustainable food source for both humans and other species (Wadrill et al. 2023). Imagine the shedding of human skin cells as a contribution to the skin of the earth. The mound transforming itself into superorganism; a city capable of regeneration. A megalopolis composed of organic matter and waste, recycled and decomposing, but breathing nevertheless. Beckett would be proud.

A similar mound of earth, spectacular root systems, and grass appear in the 2019 Red Stitch, Melbourne, production of Caryl Churchill's *Escaped Alone* (2016), directed by Australian writer/director Jenny Kemp. The design of this production, with set and costume by Dann Barber, interprets Churchill's minimalist stage directions – 'Sally's backyard. Several unmatching chairs. Maybe one's a kitchen chair' by positioning the four female performers sitting in a backyard on top of a mass of exposed earth.

In contrast to Winnie's symmetrical slope, Barber's design cuts through the earth so that directly below the green grass, we see a large cross-section of the underground world. What is particularly novel about this design is that it positions the soil at the eyeline of the audience with the human performers raised slightly higher, making their lives and conversations secondary to the underground performances. Spectators are forced to look up if they want to watch the trivial lives of the humans and their chatter; the human is literally decentred in favour of dirt and roots. Plant roots entwine themselves around domestic rubbish, broken kitchen items, vibrant trash. Meanwhile a series of staccato and elliptical dialogues between the performers traverse trivial suburban small talk, before being interrupted by increasingly apocalyptic monologues, evident in references to children poisoned by sugar, floods, contagion and fires:

> Fire broke out in ten places at once ... The blackened area was declared a separate country with zero population, zero growth and zero politics. Charred stumps were salvaged for art and biscuits.
>
> (Churchill 2016: 37)

Throughout, there is little reference to the environment as 'nature', yet climate change is inseparable from the performing climates here. For Peta Tait, the 'vibrant mood' of *Escaped Alone* both points back to the legacy of Beckett, like the optimistic nattering of Winnie that builds towards a kind of 'ontological anguish', and also marks a new phenomenological register – terror of climate collapse, in which a lack of political action leads to the destruction of the human and the nonhuman (Tait 2019: 286). As reviewer Robert Reid puts it, 'the detritus of now' builds to an absurd, towering, immanent apocalypse (2019). Borders collapse, as the 'charred stumps salvaged for art' become tangled by association with the earth below the performers' feet.

The culmination of the work is the repetition of two words:

> Terrible rage terrible rage terrible rage terrible rage terrible rage terrible rage terrible rage terrible rage terrible rage terrible rage terrible rage

terrible rage terrible rage terrible rage terrible rage terrible rage terrible rage terrible rage terrible rage terrible rage terrible rage terrible rage terrible rage terrible rage terrible rage terrible rage.

(Churchill 2016: 42)

It is a human voice that repeats this phrase, in a virtuosic escalation from veteran actor Julie Forsyth, yet the roots and earth are always present in the background, intermingling with the words, the exhalations, the movement of air and the changing theatrical climate. The blackened area – while devastated – thus brings to mind the regenerative capacity of forests, a capacity drawn upon by the traditional agricultural practices of Indigenous Australians, such as periodically burning sections of designated bushland to promote growth (Pascoe 2014: 119). This entanglement of the human and more-than-human is distinct to the Red Stitch/Kemp performance of *Escaped Alone*, wherein the potential of regeneration is also close to the surface. Compost as performance is not an apocalyptic mode but one that draws attention to the specific ground underneath our feet.

Engendering the Future

The terrestrial possibility that Latour articulates *Down to Earth* is based on the ability of beings to cultivate attachments at the very same time as they are continually moving across borders, overlapping and embedding themselves within one another (2018: 83). The process of engendering is a process of composition and composting. The key for the Earth, as for theatre as a political form, is to move beyond the label 'Made by Humans' (Latour 2018: 94) and to embrace their bacterial qualities, their ability to subvert scales, times, dimensions and to encompass an entire set of animate and nonanimate actors, all treading the boards in their own way, in their own time. This is the moment when the sphere of theatre moves towards the terrestrial earth. Winnie's mound might also be the home of an ant colony. Or a microbial galaxy.

Performance as compost is not simply a metaphor. It reveals how the 'shimmer' of life might be part of the arts of the future (Rose 2017), reaffirming the potential of performance to illuminate the interconnections between things, and to create new aesthetic and ethical entanglements with ecologies, both theatrical and profoundly political. Consider the earth of the leafcutter ants: it is a society *and* a dwelling, both air-conditioning and compost.

Notes

1 For a discussion of the 'shimmer', with reference to the work of Deborah Bird Rose (2017), see Chapter 13: Unicorn.

2 *High Life* references *Silent Running* (1972), in which the last living species from earth are housed in giant geodesic domes, greenhouses for space. However, *High Life* is also an antidote to the colonial composting regime of Mark Watney (Matt Damon) in *The Martian* (2015), a botanist marooned on Mars who promises to

'science the shit' out of his predicament by growing potatoes, or the ending of *Passengers* (2016) that celebrates tree planting (carbon off-sets?) in the hull of a spaceship. Indeed, Denis' preoccupation with the fertility of the soil stands in contrast to the devastated dust bowl corn fields and okra of *Interstellar* (2014), the doomed hydroponic oxygen garden on the spaceship Icarus II, featured in *Sunshine* (2007), and even the paltry vegetable bed of *Biosphere* (2022).

3 Despite this common name, termites are not, in fact, ants at all. Rather they are a distant descendant of cockroaches. However, they are similar to ants in that they live in large 'superorganism' colonies.

References

Beckett, Samuel (1990). *The Complete Dramatic Works*. London: Faber & Faber.

Biosphere (2022). Film, Dir. Mel Eslyn. Duplass Brothers Productions, United States.

Churchill, Caryl (2016). *Escaped Alone*. London: Nick Hern Books.

High Life (2018). Film, dir. Claire Denis. Andrew Lauren Productions/Pandora Filmproduktion/Alcatraz Films/The Apocalypse Films Company/Madants, France.

Interstellar (2014). Film, dir. Christopher Nolan. Paramount Pictures/Warner Bros. Pictures/Legendary Pictures/Syncopy/Lynda Obst Productions, United States.

Latour, Bruno (2018). *Down to Earth: Politics in the New Climatic Regime*, trans. Catherine Porter. Cambridge: Polity Press.

The Martian (2015). Film, dir. Ridley Scott. Scott Free Productions/Kinberg Genre/TSG Entertainment, United Kingdom.

Montgomery, David R (2008). *Dirt: The Erosion of Civilizations*. Berkeley: University of California Press.

Pascoe, Bruce (2014). *Dark Emu, Black Seeds: Agriculture or Accident?* Broome, WA: Magabala Books.

Passengers (2016). Film, dir. Morten Tyldum. Columbia Pictures/LStar Capital/Village Roadshow Pictures/Wanda Pictures/Original Film/Company Films/Start Motion Pictures, United States.

Phillips, Perdita (2001). *Termitaria*. Artwork.

Phillips, Perdita (2015). *Termite Embassy*. Artwork.

Reid, Robert (2019). 'The Detritus of Now: Escaped Alone'. Review for *Witness Performance*, 6 June. www.witnessperformance.com/the-detritus-of-now-escaped-alone, accessed 21 November 2019.

Rose, Deborah Bird (2017). 'Shimmer: When All You Love Is Being Trashed'. In Anna Tsing, Heather Swanson, Elaine Gan and Nils Bubandt, eds, *Arts of Living on a Damaged Planet: Ghosts and Monsters of the Anthropocene*. Minneapolis: University of Minnesota Press, pp. 51–63.

Silent Running (1972). Film, dir. Douglas Trumbull. Trumbull/Gruskoff Productions, United States.

Sunshine (2007). Film, dir. Danny Boyle. Moving Picture Company/DNA Films/UK Film Council/Ingenious Film Partners, United Kingdom.

Tait, Peta (2019). 'Striving, Falling, Performing: Phenomenologies of Mood and Apocalypse'. In Peter Eckersall and Helena Grehan, eds, *The Routledge Companion to Theatre and Politics*. London: Routledge, pp. 284–287.

Tubiana, Laurence, Bruno Latour and Frédérique Aït Touati (2015). Climate – Make It Work. *Performance Event*.

Wadrill, Violet, Topsy Dodd Ngarnjal, Leah Leaman et al. (2023). *Tamarra: A Story of Termites on Gurindji Country*. Melbourne: Hardie Grant.

6 Mermaid

We begin with mermaids or merfolk – aquatic creatures which appear in stories across global cultures. Greco-Roman merfolk or water nymphs are ancient mythical creatures that not only populate the oceans but also show up in terrestrial spaces. Found on beaches, headlands, shingles, springs and riverbanks in different hybrid combinations of human, fish and bird (Davis 2019: 257). It is the feminine form of merfolk, the mermaid, that has captured the human imagination most vividly and been the creative inspiration for many stories. From the works of Homer to the Hollywood romcom *Splash* (1984) starring Daryl Hannah, the golden mermaid Suvannamaccha in Hinduism and Thai and Cambodian versions of the Ramayana, to the Japanese animator Hayao Miyazaki's *Ponyo* (2008), the mermaid's dangerous allure as a sexualised, ship-wrecking siren, distracting otherwise level-headed captains from their duties is now a familiar trope. Yet the iconography with which we are most accustomed – the woman's torso adjoined to a fluke tail – was only consolidated in the twentieth century (Davis 2019: 258).

The Insta-Mermaid

Like unicorns, today mermaids are not only the stuff of legend but also real, enterprising and sometimes activist creatures with market and cultural influence. Sara Malou Strandvad and co-authors detail the work of performers who are 'professional mermaids' in a 'mermaid industry' (Strandvad et al. 2021a: 69, 76). This entrepreneurial practice found success in the 2000s and, since around 2012, certain subgroups of these performers use their mermaiding careers for marine advocacy (Strandvad et al. 2021b: 263). These mermaids are skilled performers, stunt workers and often professional models. They swim a fine line between entertainment business capitalism, Instagram tourism, self-promotion and fan culture that often reinforces patriarchal norms of feminine beauty and mastery over the nonhuman.[1]

Yet, since the Disneyfied mermaid cuts a glamorous, arresting figure, the Insta-mermaids' appearance in performance and activism offers a potentially powerful platform to capture the attention and intrigue of audiences. Strandvad and co-authors note that some high-profile mermaids work with NGOs to

DOI: 10.4324/9781003273080-6

be photographed or filmed swimming with wild and endangered animals and use their platforms to draw attention to ocean pollution and overfishing (2021b: 266, 272). Their interstitial or cross-species dimension invites spectators to interrogate the entanglement of human-more-than-human relations and the limits of human perception. These social media mermaids often disguise the mess of matter and climates through their mediation and the individualistic, Insta-perfect shots that make them capitalist clickbait. The watery microclimates in which they work and the ways they teach their bodies particular choreographies for the water are glossed over by their digital imprint, though they are indeed complex and nuanced, as revealed by Strandvad's analysis (2021b). The mermaids work hard to appear non-compostable, even when they are doing blue activism.

A Fish Out of Water

By contrast, in the live performance space in February 2020, a mermaid appeared in the Sydney Festival who drew sharp attention to the microclimates in which she was immersed. *The Mermaid* was written and performed by Australian artist Hanna Cormick[2] who was once an accomplished circus, dance, acrobat and physical theatre performer with a thriving artistic career in Europe. Cormick's 2020 mermaid, however, could only move slowly and carefully. Cormick wore a full-face respirator mask that completely covered the face except the ears, was attached to an oxygen tank and used a wheelchair. This mermaid was sick and suffering. While the mythical creature only exists in our imagination or on the fantasy world of the stage, the illness here was real (Figure 6.1).

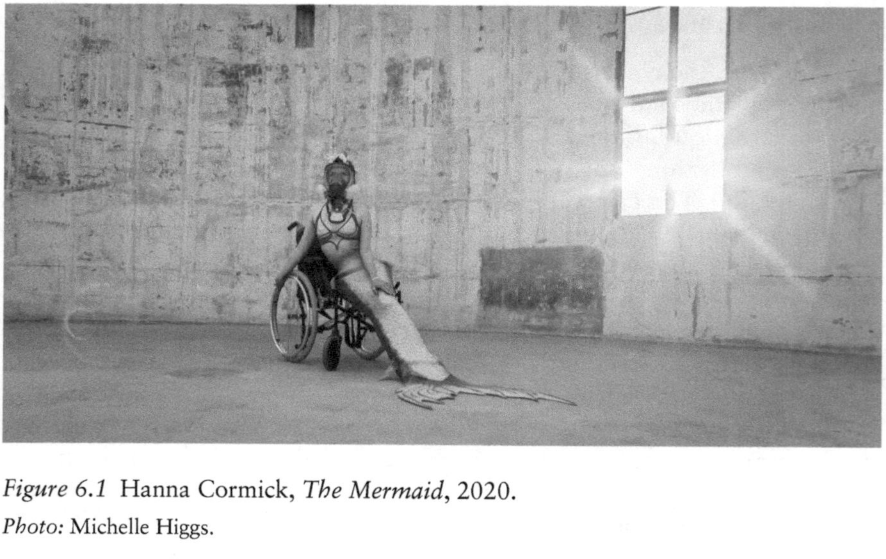

Figure 6.1 Hanna Cormick, *The Mermaid*, 2020.
Photo: Michelle Higgs.

In 2014, Cormick began to experience symptoms of several rare immune conditions which made every part of Cormick's body prone to dislocation and severely allergic to pollutants in the air: everything from food particles, washing detergents, to ink and smoke (Cormick 2021). The world had become too toxic for Cormick's body to process and Cormick was forced to live in a sealed room. Cormick came to terms with disability and illness by hiding away in shame and confusion, rarely going into public and ceasing to perform. The difficulty experienced in accessing or moving through public space made Cormick feel segregated from the rest of society, a feeling that is commonly experienced by disabled people (Cormick 2021). Cormick describes feeling invisible and silenced, not only because it was very dangerous to be exposed in public space but also by the stigma of disability which was being intensified by an increasingly polluted atmosphere and biosphere (Cormick 2021).

Reacting against this unspoken public pressure to hide a disabled body and stifle its voice, the autobiographically based personae that Cormick creates – the mermaid – is, like the Insta-mermaids, spectacularly visible. In the performance, the mermaid describes her public hypervisibility as an attempt 'to cauterise her own shame' (Cormick 2020). She wears a metallic purple-blue scaled tail and matching bikini top. Sitting above her face mask is a crown of white shells. Cormick's body is thin, white, delicate yet muscular. Signs on the walls of the performance space read, 'If you are wearing perfume, cologne, deodorant or makeup, please move to the back of the space' (Cormick 2020). The audience is informed in the programme that the solo performer suffers from a series of autoimmune conditions that restrict breathing and movement and that the toxins the spectators bring with them into the space, inside or outside of their bodies – in their stomachs, clothes or bags – are hazardous to Cormick's safety. A medical team are stationed around the performance space, armed with EpiPens in the probable case of a medical emergency.

When we go to the theatre, we rarely think about the air inside the venue, unless it is particularly hot or cold. Yet the increase in human respiratory illnesses and morbidity due to worsening air pollution was already acutely foregrounded at the time of *The Mermaid*'s first performance due to the Black Summer fires of 2019-2020. *The Mermaid* was performed in a repurposed coal loader, now a trendy arts precinct in North Sydney. While the work was originally designed to be performed outside, the smoke from bushfires forced the work indoors into a nonconventional performance venue. The climate of the performance space of The Coal Loader, with its elemental particles swirling through the air (including the particulate matter of the bushfire smoke), was rendered hypervisible by the signage, programme notes, Cormick's mask and the audience's proximity to the raging bushfires.

The visibility and vulnerability of Cormick's body and contaminated climates make *The Mermaid* a performance of disability and climate activism (see Stevens and Varney 2022). Cormick uses the figure of the mermaid, familiar to audiences through literary and popular culture, particularly through Hans Christian Andersen's 1837 children's story *The Little Mermaid* (2017) and its Disney cartoon adaptation (1989), to make the often-invisible pollution of the air and its imbrication in living bodies tangible to audiences and worthy of urgent attention. The performance connects the micro particulate pollution to macro questions of climate change and highlights the mess of matter between, through and across human and more-than-human bodies gathered within The Coal Loader.

When the performance begins, Cormick as mermaid is lying on the ground. On one side of the stage stands an Auslan interpreter who signs for the d/Deaf. We hear Cormick's laboured breathing in the mask amplified through the speakers and see the rise and fall of the mermaid's chest. The performance begins with Cormick's crisply articulated voice reading a quotation from American black feminist and activist, Audre Lorde:

> That visibility which makes us most vulnerable
> Is that which also is the source of our greatest strength.
> Because the machine will try to grind you into dust anyway
> Whether or not we speak
> We can sit in our corners
> mute forever while our sisters and our selves are wasted
> While our children are distorted and destroyed
> while our earth is poisoned
> we can sit in our safe corners, mute as bottles and we will still be no less afraid.
>
> (Lorde in Cormick 2020)

The combined challenges of female visibility, muteness and disability are central to the mermaid's most famous historic narrative and the porous boundary between life in the sea and life on land.

In Andersen's classic tale, the youngest mermaid is a fish out of water who desires an immortal soul that she is told humans possess and which she too can acquire if she convinces a man to fall in love with and marry her (2017: 280). She engages a sea witch to give her legs in exchange for her voice (2017: 283). Yet she too is rendered 'disabled' in both sea and on land. In her case, it is not due to genetics or challenging climates, but rather the traditions of her culture and the radical sacrifice she makes in pursuit of an immortal soul. Andersen's little mermaid is high born and is forced to wear eight large oysters on her tail, a painful marker of her rank. In *The Little Mermaid*, this maiming is aligned with social and class status along gender lines, where, lest she be too strong, her power is curtailed, as it were.[3]

When the little mermaid trades in her tail, the sea witch warns her that acquiring a new pair of legs will be painful too. Andersen describes the little mermaid's experience of growing legs 'as if a two-edged sword went through her delicate body' (2017: 285) and that walking was like 'treading upon the points of needles or sharp knives' (2017: 286). Since she has no voice to charm the prince, she resorts to dancing, exacerbating the pains in her limbs but making a beautiful and enchanting vision for the prince.

Unlike the little mermaid, Cormick's pain is not a voluntary sacrifice along gender and class lines. Cormick positions herself on stage as a fish out of water, only, in her case, both the land and the seas are hostile, polluted spaces in which she struggles to breathe.

Although Cormick highlights the significance of voice in the performance, the muting of other others is rendered more apparent by the performer who signs for the d/Deaf. In *The Mermaid*, Cormick connects the silencing of minority groups with the impacts of climate change and industrialisation saying:

> I am the voice of the voiceless
> I am the face of the invisible
> I am the low-lying islands we are drowning
> I am the sick air, sick ocean, contaminated water, earth
> I am the damage we have done to the earth
> I am all the people you hide away and pretend do not exist
> I am everyone you tread on to stand where you are
> I am the canary in the coalmine.
>
> (Cormick 2020)

Cormick's mermaid gives a voice to nature and stages her body as a microcosm of how much damage humans have done to human and more-than-human bodies and how much more suffering we might all endure if this continues. This mermaid does not need catchy melodies or cute dancing crab choreographies to render the mermaid's voice alluring and powerful; her words are shocking revelations about the effects of an over-industrialised world on her body and the potential for us to wake up and discover that we have made the earth uninhabitable for our fragile human bodies.

The Mermaid further illustrates how theatre spectators are not only complicit in petrochemical climates but also its active (perhaps unwitting) agents and champions. This was made particularly vivid when Cormick's body tipped into seizure during the performance (sometimes multiple seizures in a single performance), an immunologic reaction triggered by the pollutants in the air including the particles of deodorant, food, makeup and smoke brought in by the spectators. Each time this happened, one of the other performers held up large flashcards explaining what was happening to Cormick in medical terms and how much pain it inflicted. Performing climates are spelled out here, at alarming expense to the performer, inviting spectators to see the movement of petrochemical atoms through the air and their impact on Cormick's cells as well as her immune and nervous systems.

Foregrounding the atmosphere as a vibrant more-than-human space figures interestingly in Andersen's *The Little Mermaid* too. In this narrative, the air is described as 'mild and fresh' when the young siren surfaces from the water for the first time. Unlike Ariel in the Disney version, Andersen's little mermaid does not succeed in seducing her prince. Instead, he views her as an adoring and devoted child or dog-like companion (he kindly leaves a cushion at his bedroom door for her to sleep on!) (Andersen 2017: 286). The bargain she has made with the sea witch means that if he marries someone else, she will lose what remains of the 300-year life cycle of a mermaid and automatically become, like her ancestors, foam on the surface of the water. The little mermaid realises she has lost the wager as her prince lies in bed asleep with his new wife and she is given one last chance to murder him.

Instead of stabbing him with a magical dagger, the little mermaid throws herself into the sea and transforms into foam. The rising sun touches the foam and transforms her into one of the many 'daughters of the air' (Andersen 2017: 292): the ultimate self-composting. But that isn't the end of her journey. It turns out that, as an air sprite, she can also acquire an immortal soul through performing good deeds such as her capacity to 'travel to the warm climates and cool the sultry air that destroys mankind with the pestilence' (Andersen 2017: 292). Where Andersen's little mermaid becomes a purifying coolant, an air-conditioning sprite as it were, washing away the pestilence of mankind, Cormick's mermaid is a reminder of how our species is now creating its own pestilence, rendering the air dangerously toxic and even unliveable.

Beyond critiquing the quality of the atmospheric commons, Cormick's mermaid evokes the absent sea or what she describes in the performance as 'contaminated water' (Cormick 2020). *The Mermaid* invites us as spectators to ask: why has this mermaid left the oceans and what has caused her to be sick and beached on the land? Astrida Neimanis links the fact that all land species evolved from the waters to our accountability towards preserving the health of our planet's water (2016). Although our species can no longer breathe under water like mermaids, we begin our lives in watery wombs and our bodies are made up of 60% water, which creates a relational hold over our bodies, what Neimanis calls a watery kinship, our 'fishiness' (2016: 112). The white shells on Cormick's head evoke the frequent media images of bleached coral in the Great Barrier Reef, one of the natural wonders of the world, now threatened with extinction due to global warming and the acidification of the seas. A year after Cormick's performance, the Australian Government of the day, with its track-record of climate denial, was still lobbying hard to avoid the Reef's 1981 UNESCO World Heritage Listing from moving into the category of 'endangered' (Readfearn 2021).[4]

Fishy Activism

The city of Venice and its lagoon, listed as a UNESCO World Heritage site since 1987, is another contested, watery area that many scientists and environmentalists argue should be categorised as endangered[5] (Barry 2023). To draw attention to this issue, during the 2021 G20 Summit in Venice, a group of Extinction Rebellion (XR) activists performed as half-beached mermaids. They draped themselves over Venetian steps, their tails dangling in a canal, while others lay floating in the water, a school of bright blue fish playing dead (Extinction Rebellion 2021). The action used the iconography of children's fairy tales to emphasise the protesters' mourning for the futurity of our species, aquatic life and city of Venice.[6] The XR mermaids clarified the causes of the mermaid beaching using placards and melodramatic poses, a choreography of death.

Also in 2021, the sea in the Gulf of Mexico caught fire due to a neglected gas pipeline that exploded underwater. The footage that went viral on social media showed a red-fireball churning in the middle of the ocean and was frequently compared to the evil eye of Sauron from *The Lord of the Rings* (Cramer 2021). No wonder mermaids are coming onto land. Today, digital imaging technologies, stronger media reporting on environmental issues and the global networks of social media mean that the anthropogenic impact on marine environments is more exposed than ever before. Frequently now, the stages for performing the climate crisis occur in the news media, which is more inclined to replay the spectacular Hollywoodesque imagery of humanity's profound impact upon the planet or, indeed, XR's visually alluring activist happenings.

Although placards are used in the dramaturgy of *The Mermaid*, it is the quiet, contemplative moments in *The Mermaid* that set Cormick's artistic aesthetics apart from the hyperbole of the news media or the Insta-perfect shot of social media. Cormick's mermaid epitomises the messy relations of performing climates in a climate emergency by rendering visible the networks of exchange between atmospheric actants, cyclical movements of air, its exchanges in living bodies and the resulting decay and renewal. These complex exchanges make *The Mermaid* a vivid example of compost theatre, which is to say a performance 'that exposes the encounters, transformations and breakdown of cells, of organisms, of bodies, of air, of machines' (Paterson 2019: 31), a dramaturgy that connects the microelements of the biosphere to the vast sphere of climate politics in stark and foreboding terms.

Notes

1 Though it can include a significant queer component too (Strandvad et al. 2021a).
2 Cormick does not currently use any pronouns.
3 Such a tradition brings to mind Chinese practices of foot binding which also link gender and social class to a forced disabling of the female body.
4 In 2021, Britain's Royal Navy continued to detonate unexploded bombs leftover from World War II in British waters, despite having safer methods to dispose of them. This caused deafness to local whales, dolphins and porpoises, interfered with their navigational systems and led to mass coral beaching events (Carrington 2020).
5 Venice is projected to be under water by the end of the 21st century (Barry 2023).
6 The beaching of these mermaids onto land is a disturbing reminder of the appearance of whales on beaches, the frequent incapacity of humans (despite their best efforts) to return them to the sea before they cook in their own exposed blubber, as well as our ignorance as to what provokes them to venture onto shore in the first place.

References

Andersen, Hans Christian (2017). *Hans Christian Andersen's Fairytales*, ed. J.H. Stickney. Minneapolis: Lerner Publishing Group.

Barry, Colleen (2023). 'Venice and Its Lagoon Again Escape Inclusion on UNESCO List of Heritage Sites in Danger'. *AP News*, 15 September. www.apnews.com/article/venice-unesco-world-heritage-downgrade-threat-45a7e0d51069c9f06e79b69bd8bd1ed3, accessed 10 November 2023.

Carrington, Damian (2020). 'Stop Blowing Up Bombs on Sea Floor, Say Whale Campaigners'. *The Guardian*, 3 November. www.theguardian.com/environment/2020/nov/03/stop-blowing-up-bombs-on-sea-floor-say-whale-campaigners, accessed 10 November 2023.

Cormick, Hanna (2020). *The Mermaid.* Performance work, Sydney Festival, 17–19 January.

Cormick, Hanna (2021). Interviewed by Lara Stevens and Denise Varney, 26 February, Zoom.

Cramer, Maria (2021). 'Leaky Gas Pipeline Sparks an Inferno in the Gulf of Mexico'. *The New York Times*, 3 July. www.nytimes.com/2021/07/03/world/americas/eye-fire-gulf-mexico.html, accessed 10 November 2023.

Davis, Tracy C (2019). 'How Do You Know a Mermaid When You See One? How Do You See a Mermaid When You Know One?'. *Theatre Journal.* 71(3): 257–288.

Extinction Rebellion (2021). 'Nowhere Is Safe.' *Global Newsletter* (53), 16 July. rebellion.global/blog/2021/07/16/global-newsletter-53/, accessed 14 December 2023.

Neimanis, Astrida (2016). *Bodies of Water: Posthuman Feminist Phenomenology.* London: Bloomsbury.

Paterson, Eddie (2019). 'Compost and Air-Conditioning: Beyond Biospherical Performance and towards the Shimmer'. *Performance Research.* 24(8): 31–36.

Ponyo (2008). Film, dir. Hayao Miyazaki. Studio Ghibli, Japan.

Readfearn, Graham (2021). 'World Heritage Committee Agrees Not to Place Great Barrier Reef on "in Danger" List'. *The Guardian*, 24 July. www.theguardian.com/environment/2021/jul/23/world-heritage-committee-agrees-not-to-place-great-barrier-reef-on-in-danger-list, accessed 4 October 2023.

Splash (1984). Film, dir. Ron Howard. Touchstone Films, USA.

Stevens, Lara and Denise Varney (2022). 'The Climate Siren: Hanna Cormick's *The Mermaid*'. *TDR: The Drama Review.* 66(3): 107–118.

Strandvad, Sara Malou, Tracy C Davis and Megan Dunn (2021a). 'Mermaids as Market Creators: Cultural Entrepreneurship in an Emerging Practice'. *International Journal of Cultural Studies.* 25(1): 68–85.

Strandvad, Sara Malou, Tracy C Davis and Megan Dunn (2021b). 'Skills and Strategies of Activist Mermaids: From Pretty to Powerful Pictures'. *Text and Performance Quarterly.* 41(3–4): 262–282.

The Little Mermaid (1989). Film, dir. John Musker and Ron Clements. Walt Disney Pictures, USA.

Wahlquist, Calla (2020). 'Australia's Summer Bushfire Smoke Killed 445 and Put Thousands in Hospital, Inquiry Hears'. *The Guardian*, 25 May. www.theguardian.com/australia-news/2020/may/26/australias-summer-bushfire-smoke-killed-445-and-put-thousands-in-hospital-inquiry-hears, accessed 14 September 2023.

Wright, Alexis (2023). 'The Black Summer Bushfires Killed 3 Billion Animals. They Are Our Relatives; They Deserve to Be Mourned'. *The Guardian*, 1 April. www.theguardian.com/australia-news/2023/apr/01/alexis-wright-black-summer-bushfires-vigil-sydney-speech-3-billion-animals-killed-are-our-relatives-they-deserve-to-be-mourned, accessed 12 June 2023.

WWF-Australia (2020). 'Australia's 2019–2020 Bushfires: The Wildlife Toll'. wwf.org.au/blogs/australias-2019-2020-bushfires-the-wildlife-toll/, accessed 27 January 2023.

7 Granite

> Deep-time landscapes have agency; they transcend the human. And history is embedded within and inscribed on them. Rock art provides rich insights into the detail of everyday lives, including astronomical knowledge, adornment practices, animal and plant life, and rituals. The landscape features themselves hold memory for future generations.
>
> (McGrath and Rademaker 2023: 10)

We begin with magma. Many kilometres below the surface of the Earth, molten magma – containing semi-liquid rock, crystalline natural materials and chemical elements (both liquid and gas) – moves through the crust. The magma follows paths created by tectonic shifts. It eventually pools underground and begins a slow process of cooling in chambers called plutons. Over aeons, these large masses form igneous rock that is rich in quartz and potassium: granite. As the volcanic mountains above it erode away, and as the Earth shifts and moves, granite is lifted to the surface. It is coarse, heavy, almost impermeable and emerges as part of the deep history of underground climates. Igneous rocks like granite cover roughly 15% of the Earth's surface. Their presence is linked to large-scale tectonic activity in the Earth's crust, seismic vibrations and magma flows (Young 2003; National Park Service 2015; Resources Victoria 2023).

In Wooragee, where the granite is pink-hued and contains seams of gold, the deep geological time of rock is intimately linked to the deep time of Indigenous cosmology and culture. According to Western chronologies and geological systems of dating, the granite we stand upon in Wooragee was pushed to the surface during the Devonian Bindian Orogeny, approximately 420–410 million years ago (Resources Victoria 2023). However, this temporality cannot be separated from the living culture of the Dhudhuroa and Waywurru peoples, and understandings of temporality as a kind of eternal present that belongs to the 'everywhen' of the past-present-future of Indigenous time (McGrath and Rademaker 2023: 5–6).

Only a few kilometres from our house is the cultural site of Yeddonba, sacred to the Dhudhuroa and Waywurru peoples as well as the Pangarang,

DOI: 10.4324/9781003273080-7

Quat Quatta and Minjambutta clans (Explore Beechworth 2017). Visiting Yeddonba, we walk along a small gravel path and through a stand of black cypress pines (*Callitris endlicheri*). Cicadas, flies and other insects click and buzz. We cross a series of wooden viewing platforms and walkways, moving up into the mid-canopy of multiple species of eucalyptus (Yellow Box, Candlebark, Brittle Gum and Peppermint Gum), then down again until the walkway draws close to the face of a huge granite outcrop. Visible on the granite surface, protected from the weather by an overhanging shelf, are faint images, orange tracings, rock art depicting totems and animals such as a *thylacine* or Tasmanian tiger thought to have been extinct from mainland Australia for over 2000 years (Australian Museum 2021).

Yeddonba is significant as a spiritual, ceremonial, meeting place. The large granite rock formations protect the artworks but would have also provided shelter and protection for Traditional Owners. Further, as Indigenous Australian conceptions of Country remind us (Gay'wu Group of Women 2019), while the rock art exists as a sacred link to the past, it is in fact the features of the landscape itself that are animate, spiritual embodiments of a continuous, living everywhen that links together performing climates of different human and more-than-human scales and times. As such, the granite beneath our feet in Wooragee and the moss and lichen-covered boulders in the garden and paddocks around our home are as interconnected to this ongoing animate performance as the tors at Yeddonba.[1]

While distinct from Australian Indigenous cosmology and spirituality, the notion of 'living rocks' or rocks-as-performance may be a global phenomenon. As Andrew Merion Jones argues, following fieldwork in the Kilmartin region of Argyll, Scotland, the presence of 'prehistoric' rock art does not suddenly *imbue* the rocks with agency. Rather, the rocks themselves were always already *animate* and they continue to be a living performance upon which the rock art is inscribed as a dialogue between the present and deep past (2012: 79). Jones writes that material environmental conditions, the light, weather, atmosphere and the changing composition of the rock itself (fissures, cracks and undulations) are as much a part of the ongoing performance as the motifs and lines drawn by human hands (2012).

Rocks as a climate and communications technology that intertwines human and more-than-human bodies and experiences across vast scales of time is already the focus of recent performance scholarship (Schneider 2016; Schneider with Rae 2018; Hiebert 2023). Here, we argue for geology as a key part of the compost model of performance due to the ways in which rocks reveal structural, compositional, durational and animate connections with deep history. We are interested in the challenge to performance makers, thinkers, teachers, students and spectators to step outside the biosphere and instead explore the alternative cycles of dynamic re-composition found in conceptions of deep (geological) time. We suggest that working and thinking with rock as a performing climate can fundamentally alter the physical, atmospheric and temporal possibilities of Western theatre models that historically privilege human-centric time and space.

Sentient Stones

In *Sacrifice* (2022), a team of performance makers, archaeologists, data analytics, robotics and AI experts collaborated on a project that saw the creation and fabrication of the stone carapaces inspired by ancient standing stones.[2] Standing stones or menhirs are large stones, shaped and positioned vertically by humans and located in the landscape. While examples of standing stones are worldwide – Indigenous Australian stone arrangements are often in patterns or shapes such as circles, animals, boomerangs and mazes – famous Western European megalithic monuments such as Stonehenge are commonly associated with the late Neolithic and early Bronze Age from 3300 to 1200 BCE (Hadingham 1975; First Peoples State Relations 2021).

Upon entering a darkened exhibition space in the Science Gallery Melbourne, on Wurundjeri Country, participant-spectators move alongside and between several 'standing stones'. The 'stones' were constructed by master craftsman Justin Green, inspired by photos and photogrammetrically produced models of existing stones from sites in Costa Rica, Gambia, Senegal, Morocco, Poland and Mongolia. Green fabricated the models through a combination of digital and hand-made techniques resulting in 3-D forms that are shaded, coloured and limestone rendered. The size of the re-composed stones ranges from 2.4 metres (8 feet) to 30 centimetres (1 foot) high. The stones are textured, fissured, cracked and marked so that the shadows cast by a single low spotlight play unevenly across their surface.

As participant-spectators navigate the gallery space, the stones begin to move. Underneath each stone model is a robot set atop multidirectional rollers. The robot is programmed with an AI movement function that reacts to the bodies within the space by emulating 'swarming' behaviour derived from natural-world systems of distribution, herding, flocking and territorial control. This means that as we move, the stones follow, sidle beside or surround us. The larger stones can give the effect of looming menacingly or skating on ice across the floor, whereas the smaller can seem like whirring chickens gathering for food around our feet. As lead creator and theatre maker Robert Walton describes:

> Ancient standing stones and swarming robotics are two technologies you never normally think of in the same breath. But I am taking a long view of performance, ritual, and technology. They are connected because they move us from mystery to awareness. It is not initially clear how to interact with them or what knowledge they may reveal. We need to slow down, listen differently, and change the way our thoughts and bodies move.
>
> (van Vegten 2022)

In *Sacrifice*, the stones are observably animated and aware. They react to stimuli – tracking the movement of a human body through motion capture cameras – but might be also reacting to changes in atmosphere, air quality and flow, heat and light. In short, the stones are interconnected to their performance climate. Further iterations might equally employ human biometric data – such as eye tracking, sweat and heart rates – as data for the AI. As Walton suggests, the stones may provoke different registers of time, ambience and sense, a slow dramaturgy, but also an increased awareness of bodies moving differently and interactions that are vibratory and mysterious. Indeed, as technologically ambitious as this piece is, the idea of the stones being animate and 'intelligent' links readily to rock performance as a form of climate and communications technology that merges deep history within the contemporary moment.

Yet, in *Sacrifice*, the human body remains at the centre of the drama. The robots are programmed to prioritise human movement and react accordingly. Contrastingly, in Belgian-based artist Kris Verdonck's *Detail* (2019), a massive 650-kilogram marble boulder hangs from a steel cable in the ceiling of a gallery space. The boulder is rigged to a ball bearing, allowing it to fully rotate on its axis. Above, a motor acts upon a steel wheel that acts upon the ball bearing and puts the rock into motion. The motor is powered by a solar panel mounted to the gallery roof. The level of solar intensity dictates the speed at which the motor operates and thus the speed at which the boulder rotates. A level of 100% solar intensity, for example, equals 25 volts and 1.2 revolutions per minute.

In Verdonck's work, the focus on human bodies is replaced with a rock performance that draws on, is literally 'powered by', solar energy and the movements of celestial bodies in the solar system, the orbiting of the Earth and the rising and setting of the Sun. Moreover, beyond the Laws of Motion and these elemental conditions, *Detail* is also engaged with more subtle changes of weather, atmosphere, pressure and the seasons. As solar intensity fluctuates over the course of a period of daylight, the rock rotates at different speeds and force. Seasonal variation, cloud cover and hours of daylight are obvious factors. Further, changes in the climate outside the gallery space influence the atmosphere inside. The temperature in the gallery may rise as the rock rotates faster and the motor generates heat as it whirrs more quickly, air flows may alter as the boulder moves, and the conditions of light, energy and sound all change according to the rock's interconnection with planetary conditions.

A note on this installation quotes the opening sentence of Samuel Beckett's novel *Murphy* 'The sun shone, having no alternative, on nothing new' (1938). For Peter Eckersall, the dependence of *Detail* on the 'poor old sun' demonstrates the work's fragility. Even as the rock is connected and relational, beautiful and sublime, it is also a provocative expression of environmental crisis and anthropocenic change (Eckersall 2020: 89). Moreover, in the terms of Felipe Cervera, *Detail* is an example of a 'planetary' performance which reveals the Earth to be part of 'matrix of multiple co-presences' (2017: n.p.). In this work, the relationship between the Earth with the Sun at the centre of the solar system reveals the materiality of the planet 'is not just of the planet as our planet, but also as one of many planets' (Cervera 2017).

This rock performance draws attention to quality of the rock itself. In this case, the marble was sourced from a quarry in Switzerland and is a type of rock historically endemic to the region of Europe encompassing France, Switzerland and Belgium (Verdonck and van Baarle 2022). Just as the granite of Wooragee cannot be separated from the cultural knowledge and spirituality of the Dhudhuroa and Waywurru peoples, Bruno Latour recognises the *terroir* of the French vineyard as a nuanced example of how biological cycles of earth and soil are inseparable from the deep histories of place (for both humans and more-than-humans) and therefore offers a possible model of a radical politics of the future. For Latour, we need to stop speaking about humans and refer instead to '*terrestrials* (the Earthbound), thus insisting on *humus* and, yes, the *compost* included in the etymology of the word "human." ("Terrestrial" has the advantage of not specifying the species.)' (2018: 86, italics in original). In this way, the rotating marble rock of *Detail* is connected to a particular geological cycle of composition and cannot be disconnected from its material and historical entwining with the mess of multispecies matter: earth, soil, light and bacteria. As such, Verdonck's *Detail* operates at shifting scales that are both planetary and microscopic.

Seismic Scales

In tracing the durational and ambient conditions of rock performance, it is important to note both the atmospheric and material conditions that act upon the external surface of the rock are always in dialogue with the 'invisible' climatic conditions that have, over millennia, shaped and formed the igneous structure before it emerged out of the Earth's crust. In the 2015 exhibition *underground*, *Nipaluna*/Hobart-based artist Lucy Bleach collaborated with the School of Natural Sciences at the University of Tasmania, the Institute of Mine Seismology, and the construction company Unique Earth to create a single large-scale work that reacted to seismic and vibrational data. Using crushed and rammed concrete, earthmover mining vehicle tyre rubber inner tubes, hosing, a mobile seismic unit and motion sensors, Bleach constructed an arc wall of recycled concrete that was placed in the centre of a rectangular gallery space.

Embedded into the concrete were two inner tubes that were connected via hoses to an air compressor. As the exhibition description states:

> *underground* sourced global seismic data from live web-based monitoring streams, as well as local 'vibrations' pertaining to human activity collected within the gallery administration and public spaces. The two streams of vibrational data were translated to electronic impulses, triggering the air compressor to release air and inflate the corresponding inner tube. Contingent on the intensity and duration of the local or global live seismic events, the inner tubes distorted within their rammed

material constraints, shifting and destabilising the overall form, triggering a process of destruction over the course of the exhibition.

(Bleach 2015)

While the recycled concrete used in Bleach's work is not technically igneous rock, we are interested in the time-lapse video of *underground*, available through the artist's website, which reveals three distinct temporalities that overlap in this performing climate (Bleach 2015). The first timescale is, similar to *Detail*, likely connected to celestial movements, as the time lapse shows the gallery lights dimming and rising in a reflection of daylight (and gallery opening) hours (Figure 7.1).

The second timescale is that of the rock artwork itself. Documentation of *underground* largely consists of the wall shown alone in the gallery space. As the artworks draw in local and global vibrations, the concrete cracks open to reveal the two inner tubes. The black rubber forms emerge from the wall, appearing to expand and contract, sway and bulge. In this way, whereas granite and marble may take aeons to be lifted to the Earth's surface and reveal the seismic forces that shaped their formation, *underground* shows an accelerated and amplified version of underground vibrations. The concrete wall is very much 'alive' as the 'invisible' seismic data is made visible.

Ultimately the inner tubes puncture and deflate, and debris gathers on the gallery floor. Fragments of concrete, from brick-like shards to microscopic

Figure 7.1 Lucy Bleach, *underground*, Contemporary Art Tasmania, 2015.
Photo: Peter Mathew.[3]

dust particles, pile at the foot of the wall and float and circulate in the air. Some of these particles may inevitably enter the lungs and respiratory systems of the humans that enter to view the work, and these particles confirm the final time layer evident in the documentation of *underground*: the time and scale of human bodies and experience. In the time-lapse video of this artwork, human bodies appear only momentarily. They are like shadows in the space that disappear just as suddenly as they arrive; human time appears largely tangential to the vibratory data that animates this work. Interestingly, the major human interaction with *underground* is seemingly not during the public showing, but during the final clean-up and destruction of the wall. Here, the time lapse shows a small team of human bodies and wheelbarrows entering the space. The wall is demolished through the human force of sledgehammers and interactions with the mechanical, vibratory force of jackhammers that shake and break the concrete before it is removed.[4]

Tellingly, the two inner tubes usually used in giant earthmoving machinery for mining operations are shaken loose from the concrete and carried away in the final moments of the time-lapse recording of *underground*. Indeed, though *underground* does not explicitly engage with Indigenous Australian experiences of time and place, for us, this artwork brings into focus performing climates that link together different human and more-than-human scales and times, revealing continuing tensions between Western notions of geology, chronology and Indigenous conceptions of Country. In Australia, colonial practices of resource extraction, in particular mining, are an ongoing example of the devastating violence visited on Traditional Owners, their lands, culture and spirit. Notably, in landmark works such as Bangarra Dance Theatre's *Wudjang: Not the Past* (2022) and *Terrain* (2012) and Marrugeku's *Cut the Sky* (2016); Australian Indigenous performance makers have long highlighted the links between the desecration of Country, climate disaster, resource extraction and colonisation. However, as Denise Varney, Peta Tait and Lara note in their project *Towards an Australian Ecological Theatre*, settler-Australian performance, especially within mainstream theatre culture, only engages haphazardly with interlinked consequences of colonisation and climate change (2021–2024).

As we write this chapter on rocks, the Australian citizenry is about to vote in a referendum on Indigenous recognition: A Voice to Parliament that will recognise the First Peoples of Australia in the Constitution and establish a body called the Aboriginal and Torres Strait Islander Voice (National Indigenous Australians Agency 2023).[5] Along with the decades of activism that proceeded it, the recent history of the Voice is intertwined with the image of a rock, known colloquially in Australia as The Rock and correctly as Uluṟu: an ancient red sandstone rock formation vital to national iconography, tourist identity and a living cultural monolith holy to the Aṉangu people of central Australia. In 2017, at the National Constitutional Convention, Elders and representatives of the First Nations peoples of Australia shared an invitation to non-Indigenous Australians entitled the *Uluru Statement from the Heart*.

The Statement called for the establishment of an Indigenous Voice to Parliament, as well as processes of Treaty (agreement-making) and truth-telling.[6]

The Statement from the Heart draws attention to Uluṟu as the iconographic image of central Australia, the 'heart' of the country, but also makes clear that any understanding of Australian Indigenous culture and customs must foreground Aboriginal sovereignty and the ties between land, climate, soil, memory and spirituality. In the words of the Statement:

> This sovereignty is spiritual notion: the ancestral tie between the land, or 'mother nature', and the Aboriginal and Torres Strait Islander peoples who were born therefrom, remain attached thereto, and must one day return thither to be united with our ancestors. This link is the basis of the ownership of the soil, or better, of sovereignty. It has never been ceded or extinguished, and co-exists with the sovereignty of the Crown.
>
> (Uluru Statement 2023)

As we consider rock performance and the ways in which rocks can reveal alternative cycles of dynamic re-composition and temporality, alternative approaches to thinking through performance – not as a resource-extractive, human-centric endeavour but as a compostable multispecies interaction – Uluṟu and the Statement from the Heart reaffirm that ancient living cultures are the intimate knowledge holders of climates that are interlinked performances of different human and more-than-human scales and times.

Exploring the Statement from the Heart website, we find an interactive and exploratory option. Here, images and text from different Indigenous languages are overlaid on each other, and if you scroll downwards, you eventually reach an animated series of rippling lines that cross the page horizontality. Initially, the lines appear topographical, they register as hills in an undulating landscape. They might also be waves in a sea or desert dunes changing with the wind. Yet taking into account the words of the Statement and reflecting on how the First Peoples of Australia have been custodians of the land for over 65 millennia, the lines quickly take on deep time meanings. The granite formations at Yeddonba and those beneath our feet in Wooragee are not simply geographical and topographical features. This granite is alive with kinship, song and memory lines, astrological and navigational, migratory and language lines, story lines. In the seemingly hard and impenetrable surface of the granite, we share an example of living performing climates.

Notes

1 A tor refers to rock formations, typically made of granite, that emerge from the surrounding landscape following erosion or weathering.

2 The team was led by our friend and colleague Robert Walton, with Aleks Michalewics, Airlie Chapman, Goran Duric, Justin Green, Elena Vella, Daniel Williams

and Thomas Keep. A newer iteration of the work, *The Encounter*, premiered in 2024 at the Victorian College of Arts, University of Melbourne.

3 *underground* resulted from research undertaken with the Department of Seismology, Earth Sciences, University of Tasmania and in collaboration with the Institute of Mine Seismology and Unique Earth Tasmania. The work was supported by Contemporary Art Tasmania's Solo Commission programme.

4 Arguably, the last image of the time lapse is the most familiar to those of us from the theatre: a space lit from above, two cheap white plastic trestle tables pushed together, a broom leaning against the wall, a green wheelbarrow, extension cords running across the floorboards and an open door.

5 The eventual outcome of the referendum, a resounding No vote, is a stark and sickening reminder of Australia's ongoing rejection of the rights of Indigenous people and an underlying racism within Australian society and politics.

6 These processes would be undertaken by a Makarrata Commission. 'Makarrata' is a multilayered Yolŋu word understood as the coming together after a struggle (Uluru Statement 2023). For an extended discussion of the Statement from the Heart, Australian politics and theatre, see our friend Helena Grehan's work (2018).

References

Australian Museum (2021). 'Thylacine'. www.australian.museum/learn/australia-over-time/extinct-animals/the-thylacine/, accessed 9 November 2023.

Beckett, Samuel (1938). *Murphy*. London: Calder & Boyars.

Bleach, Lucy (2015). 'underground (2015)'. Work Description. www.lucybleach.com.au/lucy-bleach/selected-works/underground/, accessed 9 November 2023.

Cervera, Felipe (2017). 'The Planet as Method'. www.felipecervera.me/the-planet-as-method, accessed 9 November 2023.

Cut the Sky (2016). Performance Work, Marrugeku.

Detail (2019). Installation Work by Kris Verdonck. www.atwodogscompany.org/en/projects/detail/

Eckersall, Peter (2020). 'The Poor Old Sun: A Note on Kris Verdonck's DETAIL'. In Peter Eckersall and Kristof van Baarle, eds, *Machine Made Silence: The Art of Kris Verdonck*. Aberystwyth: Performance Research Books, pp. 89–91.

Explore Beechworth (2017). 'Yeddonba Aboriginal Cultural Site'. www.explorebeechworth.com.au/yeddonba, accessed 5 December 2023.

First Peoples State Relations (2021). 'Fact Sheet: Aboriginal Stone Arrangements'. www.firstpeoplesrelations.vic.gov.au/fact-sheet-aboriginal-stone-arrangements, accessed 9 November 2023.

Gay'wu Group of Women (2019). *Songspirals: Sharing Women's Wisdom of Country through Songlines*. Sydney: Allen & Unwin.

Grehan, Helena (2018). 'First Nations Politics in a Climate of Refusal'. *Performance Research: On Climates*. 23(3): 7–12.

Hadingham, Evan (1975). *Circles and Standing Stones: An Illustrated Exploration of Megalith Mysteries of Early Britain*. New York: Walker.

Hiebert, Ted (2023). 'Trust Fall: Rubbing Rocks and the Irreconcilability of Objects'. *Performance Research*. 27(2): 16–24.

Jones, Andrew Merion (2012). 'Living Rocks: Animacy, Performance and the Rock Art of the Kilmartin Region, Scotland'. In Andrew Cochrane and Andrew Merion Jones, eds, *Visualising the Neolithic*. Oxford: Oxbow Books, pp. 79–88.

Latour, Bruno (2018). *Down to Earth: Politics in the New Climatic Regime*, trans. Catherine Porter. Medford, MA: Polity.

McGrath, Ann and Laura Rademaker (2023). 'The Languages and Temporalities of "Everywhen" in Deep History'. In Ann McGrath, Laura Rademaker and Jakelin

Troy, eds, *Everywhen: Australia and the Language of Deep History*. Sydney: UNSW Press, pp. 1–36.
National Indigenous Australians Agency (2023). 'Referendum on an Aboriginal and Torres Strait Islander Voice'. www.niaa.gov.au/indigenous-affairs/referendum-aboriginal-and-torres-strait-islander-voice, accessed 9 November 2023.
National Park Service (2015). 'Granite and Granodiorite FAQ'. www.nps.gov/goga/learn/education/granite-and-granodiorite-faq.htm, accessed 9 November 2023.
Resources Victoria (2023). 'Victoria's Geology'. www.resources.vic.gov.au/geology-exploration/victorias-geology, accessed 9 November 2023.
Sacrifice: Contemplating Time and Trust (2022). Performance Work by Robert Walton, Aleks Michalewics. Airlie Chapman, Goran Duric, Justin Green, Elena Vella, Daniel Williams and Thomas Keep. Science Gallery Melbourne, University of Melbourne.
Schneider, Rebecca (2016). 'Extending a Hand'. Keynote Address to Performance Climate, the 22nd Performance Studies International Conference. University of Melbourne.
Schneider, Rebecca and Paul Rae (2018). 'Extending a Hand/Leading an Ear'. *Performance Research*. 23(3): 13–24.
Terrain (2012). Performance Work, Bangarra Dance Theatre.
Uluru Statement (2023). 'Uluru Statement from the Heart'. www.ulurustatement.org/the-statement/view-the-statement/, accessed 9 November 2023.
van Vegten, Megan (2022). 'Standing Stones and Swarm Robotics… Artist Dr Robert Walton Explores the Mysteries of Old and New Technology'. *Ingenium*, Faculty of Engineering and Information Technology, University of Melbourne, 30 September. https://eng.unimelb.edu.au/ingenium/standing-stones-and-swarm-robotics-artist-dr-robert-walton-explores-the-mysteries-of-old-and-new-technology, accessed 9 October 2024.
Varney, Denise, Peta Tait and Lara Stevens (2021–2024). Towards an Australian Ecological Theatre. Research Project, University of Melbourne.
Verdonck, Kris and Kristof van Baarle (2022). 'Between Machine and Prey'. Artist Talk Delivered at CLIMARTE Gallery, Richmond, Victoria, 26 May.
Wudjang: Not the Past (2022). Performance Work, Bangarra Dance Theatre.
Young, Davis (2003). *Mind Over Magma: The Story of Igneous Petrology*. Princeton: Princeton University Press.

8 Spider

We begin with a spider. A redback spider, also known as the Australian black widow. One of the world's most venomous spiders. Part of the Araneae order of arachnids (O'Hanlon 2023: 9), arthropods, animals with jointed limbs and a mineralised exoskeleton. Eight legs, many eyes of varying sizes, hairy, egg sacs, silk glands and spinnerets, webs and orbs – creators of the strongest material in the world which they pull from their backsides – beings of nightmares and of dreams. Female redbacks have a bulbous body of around ten millimetres and eight slim legs, while the males are three to four millimetres and all black. The bright red stripe on the back of the female is so vivid and artfully drawn it is hard to look away, but she is a shy, nocturnal animal and doesn't like to show off her racing stripe. Her web is less geometric than other spiders and consists of sticky and nonsticky sections. The male draws her attention by plucking on her web and performing a mating dance. As one of her common names suggests, she cannibalises the male after mating, or we might say she composts her mate – Natasha Romanoff style. The redback is one of two known species of spider where the male assists in the cannibalisation process, effectively somersaulting into her mouthparts during copulation in order to prolong the process. The red-backed spider might be dangerous to humans but, like all spiders, they perform a vital function in terrestrial food webs – keeping our environment from being overrun with cockroaches, moths and flies.

Sex and Aesthetics

Spiders have long been associated with performance and sexual passion. Think of the name tarantula which comes from the Italian town of Taranto, known as the birthplace of the Tarantella dance, a giddying, convulsive dance that was supposed to cure spider bites. Nora in Ibsen's *A Doll's House* (1879) famously danced the Tarantella in a wild and sexualised fit of passion, often associated with female sexual repression and hysteria. Nora dancing the Tarantella was the public expression of her sexual desire and rejection of the domestic sphere that held her in. Like many modernist playwrights, Ibsen anthropomorphises the animal-inspired expression to

DOI: 10.4324/9781003273080-8

make a comment about his human character of Nora, to make what was, at the time, a radical feminist social critique (Ibsen 2012). More recent performances with spiders are also interested in the relationship between the spider, sexual performance and aesthetics, though these works tend towards allowing the spiders 'speak' on their own terms and it is here that we start to come closer to a compost theatre.

In 2018, Sydney-based, Colombian-born visual and performance artist Maria Fernanda Cardoso presented an exhibition titled *The Art of Seduction.* Two of the works included in the series were 'On the Origins of Art I' and 'On the Origins of Art II': video installations created with high-definition macro cinematography and a laser vibrometer that magnified and captured the visual and audio performances of the Australian *Maratus*, a genus of eight-eyed jumping spiders (which can jump at least 40 times their body length) and are commonly known as peacock spiders (Cardoso 2018a, 2018b). These spiders – of which there are over 100 individual species – are the size of a speck to the naked human eye, but their excellent vision means they have the capacity to see the full spectrum of colour as well as the ultraviolet range. Males have bright, colourful tails with iridescent highlights like the peacock bird, while the females are a dull brown-grey and hairy (Figure 8.1). Male *Maratus* spiders use their tails to perform a kind of dance to attract a mate, like the male peacock bird or the male bird-of-paradise.

In an extraordinary mating ritual, captured on film in these performances, a male *Maratus* waves its legs like an air traffic controller on a runway, in order to attract the female's attention. Important to the dramaturgy of the performance, the *Maratus* uses his abdomen to make vibrations, creating a variety of sounds that humans describe as 'rumble-rumps', 'crunch rolls' and 'grind-revs' (Girard et al. 2011). This beat sets the pace for the dance. The vibrations and rhythm build to the point where the *Maratus* unfolds and fans out his iridescent abdominal flap and shakes it from side to side as he side-steps back and forth. If the female is impressed by the male's artistry, his beats, his colours – what might most accurately be described as his twerking – she will allow him to approach her.

In her description of the work, Cardoso writes:

> He is about 4 mm big, she is 6 mm. He is colourful and outlandish, she is not. He tries really hard to get her, she watches him. Hard to please females have driven the *Maratus* males to such extremes that they are the [*sic.*] probably the first ever performance artists: they dance and choreograph, drum and sing, all at the same time as they wave their colourful tails.
>
> (2018a)

Instead of manipulating the *Maratus* to behave in ways humans might consider interesting or aesthetically impressive, in this work, Cardoso plays the role of scientific observer and invites spectators to do the same. She offers

Figure 8.1 Maria Fernanda Cardoso, 'Origins of Art', 2016.
Photo: Maratus splendens.

a different experience to seeing the *Maratus* in nature or in a nature documentary where the spiders would be in their natural habitats. Cardoso places them against a stark white backdrop, a seemingly traditional gallery space, creating a 'stage' for the *Maratus*' dance and beats. In so doing, she makes the actions more readable to humans.

We might say that the performances were curated or stage-managed rather than created by Cardoso. She facilitates human spectatorship to an impressive more-than-human choreography, music and visual display previously invisible and inaudible to humans. She resists the temptation to overlay the *Maratus*' dance with human music (as done in popular videos of peacock spiders on YouTube). She emphasises the animal's 'complex system of courtship display that exhibits its prowess as a dancer, musician and visual artist' (Cardoso 2018b). Following Una Chaudhuri's hope that artists respect the autonomy of the animals they collaborate with in their artistic practice (2017), we might say that, in *The Art of Seduction*, Cardoso respects the *Maratus* without anthropomorphising them. In foregrounding the female *Maratus*, she accentuates the male's dance as a performance that is not *for* humans. Instead, we are privileged spectators to an intraspecies' performance, one in which we might recognise shared aesthetic values.

This, however, is not to suggest that Cardoso does not interfere with the animals or stage-manage the performance. She fully acknowledges that the mating ritual is entirely orchestrated for the camera as much as a nature documentary contains much behaviour that is not 'natural'. Such curation requires great technical precision and scientific expertise, sensitivity and ethical considerations. For this project, Cardoso collaborated with spider experts in order to locate different *Maratus* species. She applied for permits from national parks that allowed her to take away limited numbers of the arachnid from any one location. Locating the *Maratus* is painstaking work due to its size, even for spider experts. Over the years, however, Cardoso gradually learned how to recognise their movements and spider silks in the undergrowth of the bush. In order for the male to be enticed into performing the 'dance', the female *Maratus* has to be a virgin and therefore must be collected early in the season. Once the insects were in captivity, Cardoso developed techniques to capture their behaviour on film. In order to get the most ideal angles, she created a tiny stage out of paper and would brush the insects gently onto it using a paintbrush. It took her a number of years to figure out how to work with the pheromones that would prompt the desirable behaviours for filming, which in the wild tend to be released onto silk draglines and detected by chemoreceptors on the spiders' legs and palps.

We cannot know what the experience is for the intended audience, the female *Maratus*, yet there is much anticipation in waiting for the female's response to the male artistry. Since her decision to mate or not to mate is based on her appreciation of a complex combination of audio–visual–kinaesthetic aesthetics, there is a lot at stake in these dances. One reviewer of Cardoso's work with the *Maratus* writes that the spiders are 'equipped with exquisite decorative features and sensory organs so refined that a purely functional interpretation of them seems inadequate' (Goodall 2013: 53). Peacock spiders (and other animals, most obviously the peacock bird) show that animals, even those barely large enough to be seen by the naked human eye, also value beauty, that they too have an aesthetic eye and ear, that they are the sum of more than just their bodily drives. *The Art of Seduction* might be described as what philosopher Isabelle Stengers calls 'reciprocal capture', an event that captures modes of existence in which neither entity transcends the other nor forces the other to bow down (2010: 35–37). Deborah Bird Rose describes this as 'a process of encounter and transformation, not absorption, in which different ways of being and doing find interesting things to do together' (2017: G51). Cardoso piques spectators' interest in spider encounters and transformations in ways that do not exploit the drama of conflict or absorb the more-than-human behaviour into a metaphor for human relations.

Yet, it is not only the *Maratus*' visual flair – their reliance on attention-seeking choreography, spectacular colour, symmetrical pattern – that resonates with our species. It is also their use of vibratory sound as a tool: their subsonic call, which makes for a multimodal form of communication. Since spiders are thought to be the second-most sensitive creatures to vibrations (which they feel through their legs) after cockroaches, we can be certain that this performance is being experienced by its intended audience in ways that are not accessible to humans. Nevertheless, the vibratory component of the performance suggests the significance of atmospheric changes, performing climates that interact with the air and its moisture saturation and the vibrator receptors of the animal. Cardoso insists through her title that this is the 'origins of art', a radical reversal of how we humans place ourselves as cultured beings at the pinnacle of the more-than-human hierarchy. This work demands a relinquishment of dominant human behaviours and beliefs over our own exceptionalism that brings it into the realm of compost theatre.[1]

Web Tension

What is less visible to human spectators in *The Art of Seduction* are the webs and silks of the *Maratus* and the critical role they play in the mating ritual. These are the territory of Argentinian-born artist Tomás Saraceno, well known for works that destabilise the human monopoly over what counts as 'art'. Since 2007, Saraceno has been working on *Arachnophilia*, a globally collaborative, multidimensional project that aims to reverse popular public attitudes of arachnophobia. A central part of this project includes the construction and display of spiderweb sculptures which Saraceno 'creates' or, more accurately, 'curates', in collaboration with different types of spiders, with careful attention to the specificity of the species and their preferred environments. Like Cardoso, in *Arachnophilia*, Saraceno asks: 'Who is the artist? I don't know anymore' (Iscoe 2022: 23).

We saw Saraceno's *Webs of At-tent(s)ion* at the Museum of Old and New Art (MONA) in Tasmania in 2022. As we entered a pitch-black room, the gallery invigilator gruffly told us not to bump into anything as we wondered around in the dark. The only illumination to guide us were delicate lines of silvery light reflecting off the filaments of a series of large spider webs and the glow of other spectators' mobile phones. The webs were displayed in nine large glass boxes which were elevated off the ground and lit from below. Each display cabinet contained several different webs built by eight different species of spiders in total and many of the web constructions had multiple parts, each of which was constructed by a different species working in and around one another. Since the webs placed in a darkened gallery gave the impression of a cosmic cluster of galaxies, the title of each work included an astronomical name of a star or star system, the scientific name(s) of the spiders and the length of time taken for making each individual web within the assemblage.

The gestation of *Arachnophilia* began with Saraceno's curiosity about the similarities between spiderwebs and cosmic structures in, particularly in relation to the black widow spider (*Latrodectus mactans*). Out of this research developed the works: *Galaxies forming along filaments like droplets on a spider's web* (2009) and *14 Billions* (2010). While the similarities uncovered were not as straight forward as anticipated, Saraceno's web experiments are centrally about atmospheres and climates. In all these works, it is not only the spiders that perform but the webs too as vibrant, shimmering, more-than-human entities. As Sasha Engelmann puts it: 'A spider's web is the species-specific production of spacetime; it is an aesthetic as well as an evolutionary, metabolic and climatic achievement' (2016: 161). The vibrating sensitivities of the spiders mean that the cosmos is experienced differently by these creatures or that a spider 'feels the atmosphere as a dense, stirring pool' (2016: 161). *Webs of At-tent(s)ion* invites us as spectators to marvel at the 'climatic achievement' and the unique geometric innovations that respond to this dense cosmic soup without being able to experience it ourselves.

There is no doubt that Saraceno's work aims to advocate for spider rights. He often sends a performative letter to the galleries where he is exhibiting, written in the voice of and signed by the spiders/webs that champions invertebrate rights. In the letter, he asks that the curators and cleaners not wipe away webs found in the gallery more broadly: 'WE WOULD LIKE TO START BY THANKING YOU FOR RECOGNIZING OUR RIGHTS TO INHABIT AND EXHIBIT IN THIS SPACE' (Iscoe 2022: 22). At MONA, however, there was something odd and even spooky about these empty nests in their sterile, museum casings, these insect catchers devoid of life. When we asked the guard where the web artists were, he rolled his eyes and informed us that of course they were not here. This was strange given that Saraceno sees the spider/web as a whole entity and multisensory 'organ' as much as its home, a part of its body that becomes its place (Clark and Hortle 2022). This is place-based relationality at its best and something humans would benefit from reflecting upon in a climate emergency but was absent from the *Webs of At-tent(s)ion* displays. We know spiders are composters *par excellence* – believed to consume around 725 million tons of insects globally each year. Yet, by removing the webs from their natural environments and making them pristine art objects, the liveliness of the ecosystem was lost. The 'spider room' where Saraceno and his team create these works along with 300 spiders is the place where the artist spiders live, breed and die, the behind-the-scenes site where the real compost theatre takes place.

Spiderwoman

Life cycles and spiders crept into our lives when Lara was pregnant with our daughter. We had questions and doubts: should we be bringing more children into this dying world? Were we equipped to protect a daughter in particular from what this world had to throw at her? Who would her role models be? In response, Lara wrote the monologue *Not Now, Not Ever* (2019b) which was performed in the *ART + CLIMATE = CHANGE* 2019 festival when our daughter was 11 months old. Louise Bourgeois' spider mother stood over the performance, her eight spindly legs offering the protective inspiration of a ferocious mother figure. Lara performed as a hybrid between former Australian Prime Minister Julia Gillard and Marvel's Spiderwoman, superhero forces to protect the girl child in a dying, misogynistic world. Lara learned to crochet (poorly) and knitted a web on stage, lacking all the artfulness of Saraceno's collaborators.

The performance began with Lara breastfeeding the baby on stage and what could be messier than breastfeeding? In the matinee performance, she was joined by a chorus of babies sucking at breasts, bottles, toys and rattles, cooing and crying, gaaing and gooing. While breastfeeding milk to their young was thought to be exclusively the domain of mammals, Lara learned about a species of jumping spider found in Taiwan in which the mothers produce and feed milk to their spiderlings. This milk substance contains four times the protein of cow's milk and is fed to the spiderlings via the mother's egg-laying opening until they reach sexual maturity (Devlin 2018).

During *Not Now, Not Ever*, Eddie and our son came on stage for a cameo and the kids made a bloody mess of the toys while the audience laughed (and cried), thinking of or remembering their own messy familial relations. Lara spoke about the composting in her body, the sugars and the microplastics in her breastmilk, into the mouth of this new baby, the economic and political climates that devalue the women and children's labour that goes into breastfeeding and the health benefits it affords. The kids still pull Lara's Spiderwoman mask out of the costume box, put it on, puff out their chests, pump up their biceps and jump off the couch with abandon, invoking somewhere their bodily memories of the spider mother imprinted onto them.

Spinning It Together

Spiders don't have the advantage of being a charismatic species and so their significance in climates and ecological systems has long been underrated. This has not been helped by Hollywood films that emphasise their dangerous qualities and elide their necessity to our planet's survival. As James O'Hanlon writes, today we need better spider stories to incite curiosity rather than fear and the arts have the capacity to offer a renewed appreciation of these captivating minibeasts (2023: 11). By validating the idea of more-than-human artistic production in works like those by Cardoso and Saraceno, the idea of human exceptionalism is tested and opportunities for a more climate-conscious view of human-more-than-human relations is opened up. These are performances that blur once rigidly separated categories of 'nature' and 'culture' into what Donna Haraway (2003) calls 'naturecultures': a reminder that there is no existence outside of nature for any species.

When seemingly fixed definitions of what constitutes a mammal or mammalian behaviour are destabilised by a breastfeeding spider species, we are invited to see similarities between ourselves and the everyday performances of these tiny arachnids. Cardoso and Saraceno's artworks demand that we relinquish the human delusion of our monopoly over beauty, reason and aesthetic appreciation as well, more radically, as the idea that everything beautiful in this world is for human consumption alone. In all these works, there is a radical intimacy with spiders and webs, in the composting dramaturgy of the act of breastfeeding and a heightened experience of vibrations in radically new climates. There is also a challenge to face what many of us fear: spiders and the end of our mastery over earthly and cosmic systems that are more intricate than we can conceive.

Note

1 Elsewhere, Lara has described this as a model of 'deep dramaturgy', a dramaturgy that demands we surrender human exclusivity over artistic production (Stevens 2019a).

References

Cardoso, Maria Fernanda (2018a). 'On the Origins of Art I–II, 2016'. Work Description, *Museum of Contemporary Art.* www.mca.com.au/artists-works/works/2018.13/, accessed 3 November 2023.

Cardoso, Maria Fernanda (2018b). 'On the Origins of Art – Actual Size II Maratus Splendens, 2016'. Work Description, *Museum of Contemporary Art.* www.mca.com.au/artists-works/works/2018.15/, accessed 3 November 2023.

Chaudhuri, Una (2017). *The Stage Lives of Animals: Zooesis and Performance.* New York and London: Routledge.

Clark, Jane and Luke Hortle (2022). 'Spider and Stars'. Content available on the Museum of Old and New Art visitor app.

Devlin, Hannah (2018). 'Scientists Discover Spider Species That Feeds Its Young Milk'. *The Guardian*, 30 November. www.theguardian.com/world/2018/nov/29/scientists-discover-spider-species-that-feeds-its-young-milk, accessed 3 November 2023.

Engelmann, Sasha (2016). 'Social Spiders and Hybrid Webs at Studio Tomás Saraceno'. *Cultural Geographies.* 24(1): 161–169.

Girard, Madeline B, Michael M Kasumovic and Damian O Elias (2011). 'Multi-Modal Courtship in the Peacock Spider, *Maratus volans* (O.P.-Cambridge, 1874)'. *PLoS One.* 6(9): e25390.

Goodall, Jane (2013). 'Natural Aesthetics: Formations of Gender in the Work of Maria Fernanda Cardoso'. *Artlink.* 33(3): 52–55.

Haraway, Donna J. (2003). *The Companion Species Manifesto: Dogs, People, and Significant Otherness*, vol. 1. Chicago: Prickly Paradigm Press.

Ibsen, Henrik (2012). *A Doll's House*, trans. Simon Stephens. London: Methuen Drama.

Iscoe, Adam (2022). 'Arachnophile'. *The New Yorker.* 98(6): 22–23.

O'Hanlon, James (2023). *Silk & Venom: The Incredible Lives of Spiders.* Sydney: NewSouth Publishing.

Rose, Deborah Bird. 2017, 'Shimmer: When All You Love Is Being Trashed'. In Anna Tsing, Heather Swanson, Elaine Gan and Nils Bubandt, eds, *Arts of Living on a Damaged Planet: Ghosts and Monsters of the Anthropocene*. Minneapolis: University of Minnesota Press, pp. G51–G63.
Saraceno, Tomás (2009). *Galaxies Forming along Filaments Like Droplets on a Spider's Web*. Installation Work.
Saraceno, Tomás (2010). *14 Billions*. Installation Work.
Saraceno, Tomás (2022). *Webs of At-tent(s)ion*. Installation Work.
Stengers, Isabelle (2010). *Cosmopolitics*, vol. 1, trans. Robert Bononno. Minneapolis: University of Minnesota Press.
Stevens, Lara (2019a). 'Anthroposcenic Performance and the Need for "Deep Dramaturgy"'. *Performance Research*. 24(8): 89–97.
Stevens, Lara (2019b). *Not Now, Not Ever*. Performance Work.

9 Mycorrhizae

> Many people think that fungi are plants, but they are actually closer to animals. Fungi do not make their food from sunlight, as plants do. Like animals, fungi must find something to eat. Yet fungal eating is often generous: It makes worlds for others.
>
> (Lowenhaupt Tsing 2015: 137)

We begin with an underground city. In the darkness of the soil, beneath the depths of the forest floor, communities of ectomycorrhizal fungi thrive. This type of fungus is in mutual entanglement with the roots of the silver fir, *Abies alba*. The mycorrhizae reach out with thread-like hyphae that cover the outside (ecto) of the root tips in a sheath (Björkman 1970; Lowenhaupt Tsing 2015: 137). As the fine hyphae branch out, extending from the sheath, they also begin to fuse and merge with one another. This is the process of linking and stitching by which fungal hyphae form a mycelial network (Sheldrake 2020: 39). Eventually, the extramatrical mycelium makes an extensive organism within the soil and biomass of fallen needles and debris. Each tree may have multiple fungi partners and it is currently estimated that there may be between 20,000 and 25,000 species of ectomycorrhizal fungi (Rinaldi et al. 2008).

These ectomycorrhizal communities are directly descended from ancient Jurassic interfaces between fungi, plant and soil that have evolved into symbiotic partnerships.

> The term "mycorrhiza" is assembled from Greek words for "fugus" and "root"; fungi and plant roots become intimately entangled in mycorrhizal relations. Neither the fungus nor the plant can flourish without the activity of the other.
>
> (Lowenhaupt Tsing 2015: 138)

The ectomycorrhizae cover the roots and spread into the soil, helping European forest trees like silver firs, pines, oaks, as well as Indigenous

DOI: 10.4324/9781003273080-9

Australian trees such as *Agonis*, *Eucalypt* and *Acacia* take in water and nutrients.[1] In exchange, the fungi feeds on the tree system, leeching photosynthetic carbon to power its underground city (Suz et al. 2017: 370). This carbon is stored in the soil in a 'complex scaffold of dead organic material' – fungi necromass – part of an incredible cycle of carbon sequestration and storage (Frew et al. 2023).

The symbiotic relationship between the fungus and the tree relies on a cyclical dance of growth, decay and interactions between bacteria, vegetation and atmosphere, multiple life forms existing in mutually beneficial interspecies relations in a particular place and performing climate. Merlin Sheldrake argues that:

> [Fungal] hyphae are chemically irritable, responsive, excitable. It is this ability to interpret the chemical emissions of others that allows fungi to negotiate a series of complex trading relationships with trees; to knead away at stores of nutrients in the soil; to have sex; to hunt; or to fend off attackers.
>
> (2020: 45)

As we argue for composting as a regenerative model for rethinking theatrical ecologies, mycorrhizal relations offer an ancient, yet contemporary, image of evolutionary innovation that sees cellular proliferation and breakdown as a key part of interspecies exchange, connection and communication. In a dying world, theatre and performance are art forms that are uniquely placed to reveal entanglements of the human and more-than-human, yet so often our performance culture revolves around reifications of Western-centric human-centric models of thinking reliant on monocultural texts and theatrical training that struggle to accommodate difference or disability or the slow, deep dramaturgy of more-than-human time.

One common means of speculating on future performing climates is to tell and retell stories, making mess a theatrical matter, as in Sheffield-based Forced Entertainment's *Tomorrow's Parties* (2011). *Tomorrow's Parties* is coolly reserved in style and staging – two performers standing on a pallet stage framed by a single string of party lights – yet expansive in imaginative scope. Similar to other Forced Entertainment works, the performance revolves around the exigencies of competition, repetition and extended duration, a kind of nagging structure that in this case sees two performers (different members of the company have performed the work in different re-stagings) circling back to the phrase 'Or/In the Future…'. 'In the Future', says Richard Lowden opening a 2013 performance at New Art Space Amsterdam, 'there won't really be countries like there are now…'. 'Or', counters Claire Marshall, 'in the future, there won't be many people, but there will be some people…'.

For company director, Tim Etchells, *Tomorrow's Parties* captures the radical insignificance of human experience when compared with epochal scale of planetary movements.

> The performers are not experts on the future, nor students of it. Instead, like the rest of us, they are interested amateurs with a serious but small vested interest – humans whose daily existence is intrinsically bound up with the ebb and flow of bigger narratives – scientific and political, banal and larger than life – which mark the globe, sometimes transforming things slowly, sometimes shifting things suddenly.
>
> (Etchells 2011)

As the performance develops, the inevitability of catastrophic climate change becomes an inescapable element, as if no vision of human future is possible without it. 'Or', says Richard casually deadpan, 'the whole world will heat up. And the rising sea levels will lead to displaced populations and scarcity of resources…'. Slowly, as the speculations become apocalyptic and absurd, the party lights hanging behind the performers begin to dim and human time begins to alter. In near darkness, Claire quietly ends the piece 'Or, in the future', she says, 'time will speed up and people's lives will get shorter and shorter. People will be like butterflies. They'll only live for a day … In the end people will live for little more than an hour'. Can we say people are people, if they only live for an hour? And how might performing climates capture that moment of more-than-human time? In drawing a parallel between the precarious life cycle of the butterfly, colourful yet brief, *Tomorrow's Parties* looks to a future for humans that is connected to rhythms and patterns found in interspecies awareness and more-than-human cycles of birth, death, decay and rebirth.

More-than-human regenerative cycles are a key feature of much environmental art. In 1982, the Hungarian American environmental artist Agnes Denes conceived of *Tree Mountain* (1996–ongoing). This work, located at the Pinzio gravel pits near Ylojarvi, Finland and supported by the Finnish Government, takes the form of an artificially constructed elliptical mountain measuring 420 metres long, 270 metres wide and 38 metres high. Over the course of four years, in the 1990s, the former quarry was reimagined, earth for the mountain was shifted and arranged. Once there was a fertile structure, the mountain was planted with 11,000 silver fir trees by 11,000 people from all over the world (Denes 2023).

In its foregrounding of timescales extending beyond individual human lifespans – an artwork conceived of as a 400-year process – Denes' *Tree Mountain* works at the timescales of the silver fir, offering a vision of tree time and a regenerative logic based in slow growth and interspecies interaction. Yet, in the geometrical and mathematical precision of the layout of *Tree Mountain*, this artwork also cannot help but conjure an image of human intervention. Moreover, the 11,000 people included in the project were issued certificates of custodianship of a single tree (valid for more than 20 generations into the future). The tensions here between intergenerational collective care, individual custodianship and the artificial creation of what Denes calls 'the largest monument of earth' offers a glimpse into the complicated reckoning between increasing environmental awareness and the privileging of the human (2023).

In the decades since *Tree Mountain* was conceived and created, two further important elements of scientific discovery now inform our reading of this artwork, namely: the profound benefits of biodiversity and mycorrhizal systems on forest health. As Figure 9.1 shows, the initial planting of *Tree Mountain* was monocultural, a forest of silver firs laid out in a perfect pattern. While it can be assumed that over the course of 400 years, this monocultural template will be radically altered by the inclusion of other plant species, the admittedly aesthetically pleasing sequencing of the silver firs also signals the ways in which human intervention can potentially disrupt millennia-old systems of finely tuned interspecies interaction. The effects of monocultural planting are now known to have detrimental effects on the health of fungi communities underground.

Indeed, while the images of *Tree Mountain* are commonly taken from above, zoomed out, the missing dimension of this artwork is the extent to

Figure 9.1 *Tree Mountain*: A Living Time Capsule – 11,000 Trees, 11,000 People, 400 Years, 1992–1996 (420 × 270 × 28 metres). Ylojarvi, Finland.

which microscopic worlds of the mycorrhizae are flourishing beneath the surface of the forest floor, aiding the silver firs to take in nutrients and acting as a kind of amplifier for the communication network of root, soil and mycelium (Ważny 2014). As studies of pure and mixed pine forests have found, a biodiverse mix of pines, oaks and other species is far more beneficial to mycorrhizal communities than monocultural plantings (Suz et al. 2017). As research into the importance of mycorrhizae to root development and soil health has become more common public knowledge, images of tree time and slow growth must also take into account dynamic multiscalar interactions with biomass, bacteria and fungi.

On the farm, we read about trees and arboreal ecosystems as we try to decide what are the best plantings to protect our dam. In mid-winter, the bare-rooted trees begin to arrive at our place. Some have been ordered months ago and come by mail, wrapped in newspaper, sat together in a plastic bag with roots sunk in sawdust. Most of our bare-rooted trees come from a farm nearby, purpose-bred for our frosty mornings and long hot summers, and with far fewer carbon miles embedded in their travels. Indigenous Australian trees don't respond well to transplanting, so these we find at a specialist nursery as tube stock and return home with trays of trees, shrubs, climbers and groundcovers.

Instead of planting in rows, we throw a handful of stones and watch where they land to determine where each tree goes: a Dada game, a dramaturgy of wind and physics, old gardener's lore. Over three days, we plant most of them, deciduous birches to shade the tree ferns and hold the moisture near the wildlife pond, an Algerian oak as a shady firebreak, fire resistant golden ash and American maples for colour, drought-tolerant olives in the granitic banks, black walnuts to create a thick grove, candlebark gum trees leading down to the dam, cypress and wattles, and yellow box gums with sweet flowers for the bees, each group of trees intermingled among companion species (except the walnuts, that repel others), each pocket of planting for a slightly different microclimate.

Rhizomes versus arboreal systems. This overused formulation, drawn from Gilles Deleuze and Félix Guattari (2013), has followed us for most of our lives, a metaphor applied to radical theatre making and critical thought.[2] Yet, it must be said in hindsight, Deleuze and Guattari needed to dig a little deeper into the soil! If they had, different conceptions of a more messy, compostable future may have been possible. The difference in our planting this year is a small black packet of powder Eddie is carrying, which contains four species of endomycorrhizae; four species of ectomycorrhizae; four species of *Bacillus* and *Trichoderma*; and five species of beneficial bacteria. We dust each planting hole with powder and rub a little more onto the roots of the tree. After each bout of planting, we step back. The task here is to imagine the Or/future. In the future, we will see upwards, downwards and lateral growth. Or, in the future, we will see devastating drought, wind damage, or fire storms. Or, in 20 years, the trees may fill the sky and shade the ground. Or, we will die. In 50 years, our bodies may be decomposing under mature trees, thickened and gnarled in the trunk. But this image of the Or/future growth also spreads underground, as mycorrhizae reach out to multiple partners and cycle through far quicker durations of necrotic biomass, in a pattern of interconnection that will stretch from there to there to there .... As Anna Lowenhaupt Tsing (2015) puts it, fungus is a world-making bio-organism. To conceive of possible futures, we need world-making tools.

An alternative vision of multiscalar bio-organic world-building can be seen in the design and staging of Icelandic singer and environmental activist, Björk's theatrical concert *Cornucopia* (2019). Performing on a stage of four giant mushroom forms, and behind a curtain composed of shimmering mycelium-like threads, Björk and her collaborators – a choir, flautists, harpists, percussionists and electronic musicians – explore aesthetics, forms, images and sounds drawn from the more-than-human world (Pownall 2019). As acclaimed set and stage designer Chiara Stephenson describes:

> We began by looking at nature and at natural forms, and that thing of appreciating a tiny natural form and zooming in and looking at it so closely that the texture within it feels like epic landscapes in themselves ... Whether it's a piece of coral or an underground network of mycelium, the epicness of the macro of nature was something that we found quite interesting.
>
> (Pownall 2019)

The resulting stage images recall cells dividing and mutating in time with the music, as microscopic plant forms bloom and burst above or over the performers, programmed to respond to the sound produced onstage. As reviewer Benjamin Law puts it: 'Björk cries out in song and flowers bloom. Flutes tweet happily and fungal spores shudder' (2023) (Figure 9.2).

Compositionally, recent performances of *Cornucopia* draw heavily on two studio albums, *Utopia* (2017) and *Fossora* (2022). *Fossora*, which is inspired by mushrooms, mycelic webs, Netflix documentary *Fantastic Fungi* (2019) and the ability of 'underground fungal networks [to] break down pollutants and create new life from decomposing matter' (Waddoups 2022), is also explicitly connected to the regenerative cycles of plant life earlier explored by the artist in the album, app, exhibition and documentary project *Biophilia* (2011).

Similar to the multiplatform artistic and activist dimensions of *Biophilia*, *Cornucopia* moves from metaphors of mycorrhizae to the effects of catastrophic climate change on the natural world and the ways humans might act collectively on behalf of the Earth. The concert famously contains a moment when Swedish climate activist Greta Thunberg appears as a projection, directly addressing the audience and promoting collective action. While the moment of direct address might seem dislocated from the abstracted plant

Figure 9.2 Björk, *Cornucopia*, 2019.
Photo: Santiago Felipe.

forms in the stage design, Björk's music is designed to make the link between human creativity and positive interspecies interaction. 'Hope is a muscle' sings Björk on 'Atopos', the lead single from *Fossara*, 'That allows us to connect' (2022).

We are listening to *Fossora* as we travel to *A Forest for Australia* (1998–ongoing), another regenerative artistic project by Agnes Denes. In 1998, Denes curated the planting of 6000 trees of endangered species to draw attention to the threat of land erosion and desertification to Australian environments (Denes 2023). Located in the industrial suburb of Altona, 13 kilometres south-west of Melbourne's city centre on the lands of the Kurung-Jang-Balluk people of the Woiwurrung, *A Forest for Australia* is situated at City West Water's Altona Treatment Plant, only a small distance from a large petrochemical complex and former coal mining area. In a space spanning 400 × 80 metres, trees were planted in five concentric forests, with three native species featured throughout – the river red gum, drooping she-oak and the salt paperbark.

Yet, rather than thriving, the artwork was almost immediately impacted by a vision of the future that might have been drawn directly from *Tomorrow's Parties*, a decade-long, once in a millennium drought that killed half of the 6000 trees. As Clarissa Chevalier notes, the effects of climate change as well as the reliance on City West Water (now Greater Western Water), a Victorian government agency, to act as caretakers of the site 'raises complex questions about the conservation of environmental art and the future of Denes's forest' (2021: 285). Later soil testing and arboreal investigation revealed that many of the species planted were unable to cope with the high levels of salinity found on parts of the site (Greening the West 2023).

However, after a period of decline, from 2015, *A Forest for Australia* underwent significant restoration. In consultation with horticulturalists, soil chemists, landscapers and irrigation experts, and with volunteers from Altona Primary School, the artwork began to be replanted (Greening the West 2023). More recently, to commemorate the 25th anniversary of the artwork, a 2023 public arts programme 'Dear Agnes' invited 12 artists and collectives to create new works that respond to the site (https://www.instagram.com/_dear_agnes/). The resulting arts programme emphasises attempts by the local municipality, Hobsons Bay City Council and Greater Western Water to reengage with the local community, including First Nations people not previously consulted, and reinvigorate the artwork.

Our visit to *A Forest for Australia* takes place as we are in the final stages of writing this book. Although the artwork was initially conceived as an experience open to the public, *A Forest for Australia*'s proximity to an active wastewater processing plant means that we need to organise our visit far in advance and must be accompanied by a staff member from Greater Western Water. On a warm morning, a few days before Christmas in 2023, we are admitted to the site and taken on a tour of the artwork and an unexpected tour of the adjoining wastewater treatment facility.

We walk through the five circles of planting and observe that, though efforts at regenerating and caring for the trees are taking place, many of the original plantings have been decimated by the poor soil and an exposed site that is routinely impacted by drought conditions, pollution and high winds. Surviving trees, such as the river red gums (*Eucalyptus camaldulensis*) that were no doubt envisaged to grow to their typical height of between 20 and 45 metres, are still only 3–5 metres high. To supplement this lack of growth, rows of quick-growing acacias have been integrated into the artwork in an attempt to replace dead or dying trees and to offer a windbreak to those still surviving. Beneath the trees there is little groundcover planting and hardly any mid-sized or shrub planting. Similar to Denes' European projects, the mix of sizes and types of plants are restricted in favour of more uniform plantings of specific tree species.

Along with the challenging environmental conditions, the limited diversity of plant species in *A Forest for Australia* may be directly impacting the mycorrhizal networks beneath the trees. This breakdown in the compostable cycle of growth, decay and regeneration is also likely to affect the overall health of the soil and the symbiotic relation between fungal networks and trees. Recent studies suggest that mycorrhizal symbionts can help to increase plant tolerance to salinity and ameliorate some plant stress (Dastogeer et al. 2020). Indeed, due to their already high-water efficiency, Klinsukon et al. (2021) studied the positive effects of mycorrhizae on *Eucalypt* species and noted increased growth and survival under saline conditions. *A Forest for Australia* is an artwork that might directly benefit from enlivening performing climates that exist beneath the surface of the earth. Alongside further careful curation and ongoing plans for collaboration with Kurung-Jang-Balluk people and members of the local community, the regeneration of this artwork may also depend on engagement with mycorrhizal communities.

Nonetheless, following the timelines of trees that exist for many hundreds of years, *A Forest for Australia* is an artwork still in its infancy. Even so, a far more dynamic example of the interplay between microscopic organisms, waste and matter transformation can be glimpsed close by. Following our visit to *A Forest for Australia*, our guide suggests we view the artwork from a different perspective. We walk some 20 metres to the nearby wastewater facility – consisting of industrial sheds, outbuildings, pumps, equipment and several huge steel tanks with steps leading to walkways above us. We climb a series of steps and walk along the edges of three large sewerage water tanks that are bubbling and lightly churning.

Our guide explains the treatment process in which oxygen is pumped into the aeration tanks to encourage the growth of microorganisms. Bacteria develop and consume the sewage and organic matter in the wastewater. As the liquid is progressively filtered and moved across the tanks, the organic matter is broken down into H_2O and CO_2, and phosphorous is removed. At the conclusion of the process, the recycled water, now technically safe to drink,[3] is predominately used for the irrigation of parks, reserves, wetlands and golf courses in the Western suburbs.

The raised view of Denes's artwork allowed us to see more clearly the five circle 'forests', all in different stages of (re)growth. Yet, far more interesting for us was the juxtaposition of the artwork to the waste tanks. In Denes' artwork, the composting cycle of a forest – in which a diversity of plants allows decaying organic matter to be consumed by fungal networks, creating pathways for regenerative nutrients – is struggling to flourish. The bird and insect life in the forests is muted: even in early summer, there is little evidence of recent bird nests and hardly any sound and movement from insects.

In contrast, the wastewater tanks reveal not only bacterial processes that enable filtration and cleansing of human waste products on a mass scale but also interspecies connection both large and small. Wetland birds and insects flit and flicker across the tanks, moving from the recycled water to a nearby wetland, integrating the treatment facility into feeding cycles and migratory patterns. Whereas Denes' artwork seems strangely disconnected from the human and more-than-human worlds around it – the wastewater tanks are teeming with connection to a world of mess and matter.

The refuse-eating bacteria in the tanks and mycorrhizae are similar as they are both composting technologies of the *Or/future*; they intermingle with what is already there, feeding on waste and enhancing the world around them. The legacy of fungi extends from a microscopic underground city to the slow growth of forests that reach into the sky. Mycorrhizae also help soils to retain carbon by storing carbon as necromass. As an art form, theatre can also embody the Or/future. Performing climates on stage are already intimately bound by cycles of decay and regeneration, interminglings of atmosphere and matter. Theatre is an art form that produces its own unique necromass! Yet we often remain stuck replicating human-centric 'climate crisis theatre' (Watson 2022) or the straight (white) environmental play. Theatre has become a place of privileged dramas and human virtuosity from inside the biosphere. In contrast, bacteria and mycorrhizae build interspecies, multiscalar worlds beneath the surface of things to support the slow growth above. Theatre as compost needs to do the same.

Notes

1 Notably, the plant family *Proteaceae* (which includes the widespread Australian genera *Banksia*, *Grevillea* and *Hakea*) do not form mycorrhizae (Australian National Herbarium 2013).
2 *A Thousand Plateaus* was first published in 1980, right around the time Eddie was born … there's a reason we're thinking theatre through gardens and biopolitics through botany, we're just saying.
3 Victorian regulations mean that no recycled water is deemed fit for human consumption.

References

Australian National Herbarium (2013). 'Australian Fungi'. www.anbg.gov.au/fungi/mycorrhiza.html, accessed 9 November 2023.

Björk (2011). *Biophilia*. Recorded Work, App, Exhibition and Documentary Project.

Björk (2017). *Utopia*. Recorded Work. One Little Independent Records, United Kingdom.

Björk (2019). *Cornucopia*. Theatrical Concert.

Björk (2022). *Fossora*. Recorded Work. One Little Independent Records, United Kingdom.

Björkman, E (1970). 'Forest Tree Mycorrhiza: The Conditions for Its Formation and the Significance for Tree Growth and Afforestation'. *Plant Soil*. 32: 589–610.

Chevalier, Clarissa (2021). 'Eco-Phenomenology and the Maintenance of Eco Art: Agnes Denes's *A Forest for Australia*'. *Australian and New Zealand Journal of Art*. 21(2): 273–290.

Dastogeer, KMG et al. (2020). 'Plant Salinity Tolerance Conferred by Arbuscular Mycorrhizal Fungi and Associated Mechanisms: A Meta-Analysis'. *Frontiers in Plant Science*. 11: 588550.

Deleuze, Gilles and Félix Guattari (2013). *A Thousand Plateaus: Capitalism and Schizophrenia*, trans. Brian Massumi. London: Bloomsbury.

Denes, Agnes (1996–ongoing). *Tree Mountain*. Artwork, Ylojarvi, Finland.

Denes, Agnes (1998–ongoing). *A Forest for Australia*. Artwork, Altona, Australia.

Denes, Agnes (2023). Personal Website. www.agnesdenesstudio.com, accessed 9 November 2023.

Etchells, Tim (2011). '"Tomorrow's Parties" Programme Note'. *Forced Entertainment*. www.forcedentertainment.com/tomorrows-parties-programme-note-by-tim-etchells/, accessed 15 December 2023.

Fantastic Fungi (2019). Documentary Film, dir. Louie Schwartzberg. Moving Art, United States.

Forced Entertainment (2011). *Tomorrow's Parties*. Performance Work.

Frew, Adam et al. (2023). 'Hidden Carbon: Fungi and Their 'Necromass' Absorb One-Third of the Carbon Emitted by Burning Fossil Fuels Every Year'. *The Conversation*, 6 June. www.theconversation.com/hidden-carbon-fungi-and-their-necromass-absorb-one-third-of-the-carbon-emitted-by-burning-fossil-fuels-every-year-206674, accessed 9 November 2023.

Greening the West (2023). 'A Forest for Australia – Altona Treatment Plant'. www.greeningthewest.org.au/projects/forest-australia-altona-treatment-plant/, accessed 9 November 2023.

Klinsukon, C, S Lumyong, TW Kuyper et al. (2021). 'Colonization by Arbuscular Mycorrhizal Fungi Improves Salinity Tolerance of Eucalyptus (*Eucalyptus camaldulensis*) seedlings'. *Scientific Reports*. 11: 4362.

Law, Benjamin (2023). 'Björk Cornucopia Review – an Electrifying Pop Concert, Art Installation and Opening Ceremony Rolled into One'. *The Guardian*, 4 March. www.theguardian.com/music/2023/mar/04/bjork-cornucopia-review-an-electrifying-pop-concert-art-installation-and-opening-ceremony-rolled-into-one, accessed 9 November 2023.

Pownall, Augusta (2019). 'Designers of Björk's Cornucopia Show "Ignored the Concert Rule Book"'. *De Zeen*, 13 December. www.dezeen.com/2019/12/13/bjork-cornucopia-chiara-stephenson-set-design/, accessed 9 November 2023.

Rinaldi, AC, O Comandini and TW Kuyper (2008). 'Ectomycorrhizal Fungal Diversity: Separating the Wheat from the Chaff'. *Fungal Diversity*. 33: 1–45.

Sheldrake, Merlin (2020). *Entangled Life: How Fungi Make Our Worlds, Change Our Minds and Shape Our Futures*. New York: Random House.

Suz, LM, S Kallow, K Reed et al. (2017). 'Pine Mycorrhizal Communities in Pure and Mixed Pine-Oak Forests: Abiotic Environment Trumps Neighboring Oak Host Effects'. *Forest Ecology and Management*. 406: 370–380.

Tsing, Anna Lowenhaupt (2015). *The Mushroom at the End of the World: On the Possibility of Life in Capitalist Ruins*. Princeton, NJ: Princeton University Press.

Waddoups, Ryan (2022). 'Björk's New Album Is an Ode to Mushrooms'. *Surface Magazine*, 3 October. www.surfacemag.com/articles/bjork-fossora-album-cover/, accessed 9 November 2023.

Watson, A (2022). 'Contemporary Catastrophes: 2010s British Climate Crisis Theatre and Performativity'. *Contemporary Theatre Review*. 32(2): 140–161.

Ważny, R (2014). 'Ectomycorrhizal Communities Associated with Silver Fir Seedlings (*Abies alba* Mill.) Differ Largely in Mature Silver Fir Stands and in Scots Pine Forecrops'. *Annals of Forest Science*. 71: 801–810.

10 Lego

We begin with Lego, a globally adored Danish construction toy produced in bright colours and made from acrylonitrile butadiene styrene (ABS), a thermoplastic amorphous polymer. Acrylonitrile is a synthetic polymer produced from propylene and ammonia, butadiene a by-product of ethylene and styrene from the dehydrogenation of ethyl benzene. Together these compounds produce a resin that is very malleable when heated and rigid when set (Lego Group 2024). ABS is an opaque cream when first produced. To create Lego pieces, the polymer is first dyed and sorted into chunks of colour. At the Lego factory, the coloured ABS is heated to 230 degrees and then injected into metal moulding machines. It is then cooled and spat out as the familiar Lego bricks within a matter of seconds. On another assembly line, details are printed onto the more individualised Lego pieces: faces, clothing, cockpits and branded logos. Robots work to ensure that each set is filled with the correct number of pieces, large and small. Once sorted, these glossy, colourful and tough parts are boxed up and shipped off.

Lego produces 19 billion pieces a year and the company has been producing its bricks since 1949 (Lego Group 2024). As a synthetic polymer, Lego is a mass-produced form of plastic made from coal and oil and thus lies at the heart of extractive industries and climate change. The first synthetic plastic derived from fossil fuels was produced in 1907 as Bakelite and used for manufacturing mass media and communication devices (Davis 2022: 27). It was soon followed by other polymer configurations designed for warfare, including now infamous materials such as Teflon which coats military tanks. In 2021, Lego embarked on a project to make their bricks from recycled drink bottles, which was abandoned soon after because it would have meant higher carbon emissions over the products' lifetime, though how these calculations are arrived at have not been made public (Bartholomew 2023).

Lego not only covers our living room floor and swirls around the Great Pacific Garbage Patch, it is a reminder of the ubiquity of plastic on our planet. Not specifically these over-saturated pretty plastic blocks but more insidious, tiny fragments less than a millimetre in diameter shaved off of our tyres, water bottles, artificial turf, clothes, toothbrushes (and the list goes on). From the Mariana Trench to the top of Mount Everest, microplastics float in our

DOI: 10.4324/9781003273080-10

waterways and into our drinking water or are washed into the soil, sucked up into the vegetables that end up on our tables. In the sea, they float into the gills of the fish we consume, airborne in the wind they fly into our lungs. The relentless pace of human plastic accumulation and microplastic waste shows no sign of slowing. Instead, plastics are here to stay, integrated into all human, more-than-human organic bodies and rock sediment alike (Figure 10.1).

Plastics industries have been and continue to be massively profitable and desirable because plastic anticipates its disposability (Davis 2022: 32). When plastic was invented, its malleability and mouldability was a treasured

Figure 10.1 A Japanese spider crab sits atop a Lego skyscraper in Miniland New York City at Legoland, Carlsbad, California, 2012.

Courtesy: Getty Images.

attribute and these qualities, combined with its cheapness to produce, promised a democratising of culture. After years of carefully washing and sorting our bottles into bins with different coloured plastic lids, agonising over which type of container belonged where, the myth of plastic 'recycling' in Australia is exposed. As nations such as China refuse to any longer be a dumping ground for Western nations, and Indigenous communities continue to protest the contamination of their sacred Land/Country, plastic is understood as convenient but problematic.

In *Plastic Matter*, Heather Davis writes that the philosophies and ideologies that facilitated the development and global uptake of plastic without governmental regulation or accountability reveal the human anxiety with our imbrication in the natural world. Plastic, she argues, 'is a desire for an existence that is less messy, less enmeshed with earthly life ... animated by a will to forget and disavow the violence of extraction' (Davis 2022: 11). As a group of materials that are notoriously difficult to compost (or even recycle), this chapter considers performances that interrogate the ironies of plastics' promise of a less 'messy' existence, what this fossil fuel-based material manifests when enmeshed with living bodies and the future of 'the plastic arts'.

Eating Our Future

Our seven-year-old comes home from school begging to watch a YouTube show called *Lego Breakfast*, a surrealist Japanese marvel that uses stop animation to create a variety of foods out of Lego. The Lego 'foods' are placed in a kitchen environment filled with chopping boards, knives, pans, stove top gas hobs as well as real food, all deployed in a breakfast 'cooking' show: fried eggs, scrambled eggs, pancakes with maple syrup, Nutella sandwiches and waffles. A real kiwi fruit lies on the chopping board. The knife 'slices' through it and we see a cross-section of its interior as a meticulously constructed Lego replica, from the brown exterior skin to the transparent green interior and the seedy, white centre. A fish loses its Lego head before it is thrown into the frying pan accompanied by sounds of sizzling and spitting oil. A real head of lettuce is sliced up and each piece turns into Lego-constructed leaves: on the one side, the organic vegetable and the other, 'prepared' side, the plastic construction. A real life jar of Nutella pours the chocolate sauce over the Lego pancake, the brown liquid transforming into rectangular bricks which stick to the hard, angular, dotted surface of the replica breakfast food.

The surreal artfulness of the show is mesmerising, its zany logic combined with painstaking attention to detail. But it also points to a deeply sinister reality at the heart of our culture: the saturation of our food chain in plastic, or more specifically, microplastics which are lodged within all the food we eat daily. It is now well understood that the average person ingests around a credit card's worth of plastic every week (Advisors et al. 2019). Microplastics circulate through our bloodstream, sit in our lungs and penetrate cell walls. For those of us who are mothers, they circulate in our placentas and breast milk, where they are fed to our offspring which can then affect either them and/or their offspring. The image of the credit card as a measure of our cumulative plastic consumption invites us to see more clearly the relationship between capitalism, consumerism, the ubiquity of plastic and what its ingestion means for our species and other more-than-human lives. As a performance of a cooking show that lacks a charismatic host or demonstrator, *Lego Breakfast* reminds us that there is no separation between nature and culture with the real and fake happily coexisting in dialectical relation. With its bright, primary-coloured fake and organic foods, *Lego Breakfast* restores a certain sheen to the horror of what we are consuming, while simultaneously inviting reflection on the relationship between the synthetic and those organic materials which we depend upon to survive.

Over more than a century, the toxicity of plastic has demonstrated its sinister impact on living bodies. The rise of rare cancers such as angiosarcomas have been documented when toxic chemicals seeped into the water and air from polyvinyl chloride (PVC) manufacturing plants. It is no coincidence that these plants were located in poor, predominantly black neighbourhoods in Southern Louisiana known as 'cancer alley' (Davis 2022: 35). While these health impacts were once easier for the mostly white upper classes to ignore through their reduced exposure to plastic toxicity, the ubiquity of plastic pollution today is changing public awareness and political responsiveness to the issue. Even today, however, politicians often continue to gloss over the ways in which chemical dumping not only affects the health of these 'sites' and the people who live on and off them, perpetuating a silence over pollution's impact on local identity, spiritual health, movement, access to food, kinship and family ties (see Hoover 2017).

Bisphenol A (BPA) and polychlorinated biphenyls (PBCs) have been known since the 1930s to mimic the hormone oestrogen and thereby impact the endocrinal system, causing ovarian disorders, reduced sperm counts, weaker eggs, higher rates of miscarriage, testicular cancer and genital abnormalities as well as type 2 diabetes, heart disease and obesity (Chiang et al. 2017; Rancière et al. 2019; Liang et al. 2020; Stukenborg et al. 2021; Ingegnoli et al. 2022; Milanović et al. 2023). When we had our first child in 2016, we watched most of the other kids around us sucking on spill-free sippy cups (let's reduce the mess of babies everyone!). Lara feared that poison might be added to her tea by other overtired mothers during playgroup to when she talked about the advantages of washable nappies. From the moment they are born, we wrap kids in plastic and feed them micropolymers; this is normalised among all classes of atomised and exhausted parents trying to stay afloat without the village to raise the child.

Although there has been a movement away from BPA in drink bottles and food containers in the past decade, with some countries outright banning these polymers, they are usually replaced by other endocrine-disrupting chemicals. While the research catches up to prove the harm of the latest microparticles, we continue to consume toxic products. Microplastics interfere with our gut biome and their hormone-disrupting properties change our moods and impact our brain in ways that science has not yet fully nutted out due to the complex interactions and exchanges of the gut-brain-endocrinal systems. And so, when we watch a theatrical performance with human bodies we are already watching more-than-human encounters taking place on stage, we don't have to be Orlan or the Kardashians to already be plastified.

There are many performances and visual artworks that use found plastics as a mode for critiquing and drawing awareness to the saturation of plastics in our lives. In certain geographical and social contexts, these works are more relevant than others. Interdisciplinary plastics scientist of Canadian Métis descent, Max Liboiron's book *Pollution is Colonialism*, expanding the work of Michelle Murphy (2018), details the ways in which plastic pollution and waste by settlers has meant the dispossession of First Nations peoples across the globe. Liboiron points out that waste disposal always assumes a form of *terra nullius* and that colonialism used and continues to liberally use the lands of Indigenous people for their waste disposal without the consent of Traditional Owners (2021). In the report 'Violence on the Land, Violence on Our Bodies', The Native Youth Sexual Health Network details the effects of what they call *environmental violence* (Liboiron 2021: 87). Liboiron notes that this violence is both material toxicity and cultural genocide, the latter being no less devastating to First Nations populations (2021: 107).

In the Philippines, first colonised by Spain and later by the United States, waste colonialism is more present due to the islands' tropical weather conditions and geography. It is particularly evident during frequent cyclones and ocean weather events that wash plastics onshore. A Philippines applied theatre group began by producing *Oh! My Gulay!* (2010) (meaning *Oh! My Vegetable!*) as a way to promote organic and sustainable farming and became Cope/with/Land Theatre Company by 2010. In 2011, they created *Gulantang! A Children's Musical* (2011) (with book and lyrics by Jon Lazam and music by Angelica Dayao), a performance about a boy trying to protect a real-life recently discovered endemic rodent (Maiquez et al. 2018). Set designer Kulay Labitigan and director Dennis Gupa built a set from plastic 7 Up bottles collected from the local beach, turned them into a replica of a river and waterfalls of Mount Banahaw and Mount San Cristobal. Other waste materials such as newspapers, tyres, metal and e-waste, including discarded television sets, were central to the mise-en-scène (Maiquez et al. 2018).

The set was both familiar and estranged, drawing attention to the ecological peril of the local mountains, rivers and all the human and more-than-human life that depends upon them for survival. While there are countless interesting examples of artists employing recycled plastic waste materials in their work, the Cope/with/Land example is made for an audience that is already familiar with the problem which stems from the nonconsensual environmental violence of colonial practices and the Americanisation of the local culture.

Filipino theatremakers Reagan Maiquez, Dennis Gupa and Kulay Labitigan argue that theatre is not only a place for imagining possible climate solutions or even critiquing related issues such as plastics pollution. Instead, their work with Cope/with/Land Theatre Company literally explores what it means to cope with the given colonial legacies and environmental conditions with which they are burdened. They write:

> Theatre, as an assembly/assemblage, a process, a creative and critical composition, a place of actants amassing, is not only a representational and whimsical planet but an actual planetary location where we can consider and address the dire ecological issues that we are all confronting. More than this, theatre and performance is a place and a time where/when interventions against ecological change and destruction can be realized (such as recycling, education, or community-building), or, in and of itself, theatre contributes to the environmental problem as unsustainable practices are not excluded from the production of art in this period of late capitalism.
>
> (Maiquez et al. 2018)

Davis notes that art in the twentieth and twenty-first centuries is deeply imbricated with petrochemicals.[1] Photography and film were invented using different plastics developed in the twentieth century. Today digital art, its online display and storage are dependent on plastic computers and millions of miles of plastic insulation on cables, fans and warehouses – the lightness of 'cloud' communication depends on a vast and highly plasticised material infrastructure (Davis 2022: 70; Diamanti 2023: 65).

Our thinking on the relationship between art and plastic was first prompted by a conversation in 2016 when Eddie coordinated the Performance Studies international conference at the University of Melbourne. Eddie and the conference team invited as a keynote speaker the much admired and, today, greatly missed Bruno Latour who accepted the invitation. Lara played chauffeur to pick Latour up from the airport because she speaks French. Excited to meet such an important thinker, Lara had a lot of questions for the eminent philosopher (because, surely, he would appreciate these after a 24-hour flight from Europe?). Instead, the moment he got into the car, to our surprise for someone about to keynote the largest Performance Studies conference in the world, he asked us: 'What is Performance Studies?'. At first, we thought this might be a way to initiate a drawn-out philosophical conversation on the slippery nature of the discipline, one which has been experimented with by everyone from Richard Schechner, Victor Turner, Diana Taylor, Judith Butler and many more who have stretched the boundaries of its large embrace. But no, Latour just wanted to know how to translate Performance Studies into French. Lara replied: '*les arts plastiques*'.

This term has always been an awkward one since it originally applied to visual arts that were physically mouldable such as ceramics, sculpture and painting. Only more recently has *les arts plastiques* been the term used to describe the category of human-centred performance art of the 1960s onwards. We spent the car ride talking about this strange term, the plasticity of performance art and the variety of forms that it takes. In the time of what Chia-ju Chang calls the Plasticocene, thinking through the future of *les arts plastiques* invites more questions (2016: 100). Do we still value these flexible and protean qualities of the plastic arts? Or are the arts cheapened, made into yet another reproducible, easily disposable product, too often tied to the fossil fuel companies that frequently fund major theatres and galleries?[2] How should we think about performance now we know that human performers are transcorporeal bodies (Alaimo 2010), hybridised with plastic?

Australian playwright and provocateur author of *Kill Climate Deniers* (2019), David Finnigan says that artists should stop writing apocalyptic pieces about how plastics are killing us. Instead, he implores us to write romances to our love affair with what he describes as the Real Housewives of the Great Pacific Garbage Patch:

> We were charmed by plastic, we fell in love with it. Then we discovered its flaws, and we recoiled from it. Now we need to learn to come to terms with it, to find some way to accept it. After all, plastic is a part of who we are. It's in our blood, and you can't recoil from your own blood.
>
> To accept plastic doesn't mean giving up on reducing the harm it causes. Loving something doesn't mean you can't criticise it. But when you look at everyone you love, from the infant in your arms to your partner of 50 years, you have to accept that that person is part plastic.
>
> (Finnigan 2023)

As Finnigan notes, like it or not, the plastic arts have literally grown into their name. Instead of embracing this new part of our multispecies identity, an act that continues to centre the human and its human-made hybrids, we suggest that what is more interesting still is the more-than-human's creative play with the conditions of the Plasticocene.

In 2001, a group of bacteria that eat plastics, *Ideonella sakaiensis*, were discovered by scientists in a rubbish dump in Japan (Buranyi 2023).[3] The scientists who discovered it note that it was not just one enzyme doing the work of breaking down the plastic but a whole chorus of bacteria cooperating together in ways that are far from being understood. The bacteria were using their enzymes to consume the plastic, turning plastics into their raw materials, a true composting. *Ideonella sakaiensis* eats too slowly to be of any help to the scale of our current quandary but play with its DNA and we might have a substance that eats away at our problem. Alternately, it could be the beginning of another Frankenstein, something that, when released into noncontrolled spaces starts to destroy the material infrastructure of our cities and towns – then we'll really understand how much we've built our 'civilisations' on plastic.

These are theatres that we argue are worth watching with wonder and respect. For all human creativity and cleverness, we have not yet found an artificial way to compost ourselves out of the problem of plastic. The more-than-human world with its limitless adaptive capacities, already has. It is worth spending more time pondering and staging the creativity of bacteria in a world that thrives and adapts in spite of us and will continue to thrive after our species has gone. Today we are being forced to reconcile with the toxic plasticity of our planet, our '*arts plastiques*' and the intergenerational impacts of these changes. Davis notes that when plastics were invented, they promised 'freedom from the obligations and messiness of the earth' (2022: 23). Now our species' survival relies on a return to mess, a renewed engagement and acknowledgement of what Val Plumwood calls our 'radical dependency' upon the nonhuman (2002: 4).

Notes

1 In 1999, Kodak was ranked New York State's leading producer of airborne carcinogens and waterborne developmental toxicants (Niman 2003).
2 For an excellent example of this, see artist and Victorian Greens elected Member of Parliament Gabrielle de Vietri's *Maps of Gratitude, Cones of Silence and Lumps of Coal* (2019), an interactive map which displays and ranks the connections between fossil fuel industries and the arts in Australia (De Vietri 2019).
3 There are strands of bacteria in mealworms that can digest Styrofoam, wax worms can degrade polyethylene and the fungus *Pestalotiopsis* microspores can biodegrade polyurethane (Davis 2022: 85).

References

Advisors, Dalberg, Wijnand de Wit and Nathan Bigaud (2019). *No Plastic in Nature: Assessing Plastic Ingestion from Nature to People.* Gland, Switzerland: World Wildlife Fund for Nature.

Alaimo, Stacy (2010). *Bodily Natures: Science, Environment, and the Material Self.* Bloomington: Indiana University Press.

Bartholomew, Jeff (2023). 'Lego Abandons Effort to Make Bricks from Recycled Plastic Bottles'. *The Guardian*, 25 September. www.theguardian.com/lifeandstyle/2023/sep/24/lego-abandons-effort-to-make-bricks-from-recycled-plastic-bottles, accessed 8 December 2023.

Buranyi, Stephen (2023). '"We Are Just Getting Started": The Plastic-Eating Bacteria That Could Change the World'. *The Guardian*, 28 September. www.theguardian.com/environment/2023/sep/28/plastic-eating-bacteria-enzyme-recycling-waste, accessed 8 December 2023.

Chang, Chia-ju (2016). 'Wasted Humans and Garbage Animals: Deadly Transcorporeality and Documentary Activism'. In Rayson K Alex and S Susan Deborah, eds, *Ecodocumentaries: Critical Essays.* London: Palgrave, pp. 95–114.

Chiang, C, S Mahalingam and JA Flaws (2017). 'Environmental Contaminants Affecting Fertility and Somatic Health'. *Seminars in Reproductive Medicine.* 35(3): 241–249.

Davis, Heather (2022). *Plastic Matter*. Durham, NC: Duke University Press.
De Vietri, Gabrielle (2019). *Maps of Gratitude, Cones of Silence and Lumps of Coal*. Interactive Work. www.gabrielledevietri.net/maps-of-gratitude
Diamanti, Jeff (2023). *Climate and Capital in the Age of Petroleum: Locating Terminal Landscapes*. London: Bloomsbury.
Finnigan, David (2019). *Kill Climate Deniers*. London: Oberon Books.
Finnigan, David (2023). 'Falling in Love with Microplastic'. Personal Website, 7 May. davidfinnigan.substack.com/p/falling-in-love-with-microplastic, accessed 7 November 2023.
Gulantang! A Children's Musical (2011). Performance Work. Cope/with/Land Theatre Company.
Hoover, Elizabeth (2017). *The River Is in Us: Fighting Toxics in a Mohawk Community*. Minneapolis: University of Minnesota Press.
Ingegnoli, Vittorio, Lombardo Francesco and Torre Giuseppe La, eds (2022). *Environmental Alteration Leads to Human Disease: A Planetary Health Approach*. Cham, Switzerland: Springer.
Lego Breakfast (2020). Video by tomosteen, 17 July. www.youtube.com/watch?v=1RMKR8h03iw, accessed 16 October 2023.
Lego Group (2024). 'How Are LEGO Toys Made?'. Webpage accessed 12 February 2024. https://www.lego.com/ms-my/service/help/fun-for-fans/behind-the-scenes/brick-facts/how-are-lego-toys-made-blt0cb8fe682c26474e?locale=ms-my
Liang, F, X Huo, W Wang et al. (2020). 'Association of Bisphenol A or Bisphenol S Exposure with Oxidative Stress and Immune Disturbance among Unexplained Recurrent Spontaneous Abortion Women'. *Chemosphere*. 257: 127035.
Liboiron, Max (2021). *Pollution Is Colonialism*. Durham, NC: Duke University Press.
Maiquez, Reagan, Dennis Gupa and Kulay Labitigan (2018). 'Between Theatre and the Environment: The Experience of Cope/With/Land Theatre Co'. *Global Performance Studies*. 1(2).
Milanović, M, N Milošević, N Milić et al. (2023). 'Food Contaminants and Potential Risk of Diabetes Development: A Narrative Review'. *World Journal of Diabetes*. 14(6): 705–723.
Murphy, Michelle (2018). 'Against Population, towards Alterlife'. In Adele E Clarke and Donna Haraway, eds, *Making Kin Not Population*. Chicago: Prickly Paradigm Press, pp. 101–124.
Niman, Michael I (2003). 'Kodak's Toxic Moments'. *Alternet*, May 28. www.alternet.org/2003/05/kodaks_toxic_moments, accessed 6 October 2023.
Oh! My Gulay*!* (2010). Performance Work. Cope/with/Land Theatre Company.
Plumwood, Val (2002). *Environmental Culture: The Ecological Crisis of Reason*. London: Routledge.
Rancière, F, J Botton, R Slama et al. (2019). 'Exposure to Bisphenol A and Bisphenol S and Incident Type 2 Diabetes: A Case-Cohort Study in the French Cohort D.E.S.I.R.'. *Environmental Health Perspectives*. 127(10): 107013.
Stukenborg, JB, RT Mitchell and O Söder (2021). 'Endocrine Disruptors and the Male Reproductive System'. *Best Practice & Research: Clinical Endocrinology & Metabolism*. 35(5): 101567.

11 Ice

We begin with ice. The process of freezing water requires a change in molecular energy and form. The shift from liquid water to solid ice occurs by lowering the temperature of H_2O molecules resulting in an increase of hydrogen bonding within the molecule such that it forms a crystalline structure. When temperatures drop to zero degrees Celsius and below, the hydrogen and oxygen molecules in water that were moving quickly in its liquid form begin to drastically slow their energy and movement. Though the molecules continue to move even when frozen, they begin to stick together. The molecules gather and form the solid crystalline structure that we know as ice.

Ice formation requires an ice nuclei or crystal seed, an impurity such as a tiny solid particle in the water around which an ice crystal can grow. While snow is made of individual ice crystals that grow often in the atmosphere, mostly in clouds, when they reach the ground and their needles begin to melt, they compact, transform and merge into an agglomerate of ice crystals that eventually become a glacier. Gradually this compacted ice melts and returns to a liquid water state and is then evaporated back into the atmosphere and the cycle continues. Ice is a shape-shifter moving between vapour whisps, delicate, unique crystals, hard masses and wet droplets – thermodynamic phases, of ceaseless de- and re-composition, constant animation. As we note in Chapter 2, ice can decelerate rates of decay in living matter, slow the transmission of infection and growth of bacteria and, when compacted over thousands of years, trap ancient particles, objects and gases within it.

Iceblocks and Icebergs

As natural wonders that are located in places geographically isolated from most of the world's populations, icebergs occupy a unique place in the human imaginary. In humans, icebergs have inspired the awe of the sublime – described as terrifying, amazing, overwhelming, exhilarating, spiritually enlightening, hallucinogenic, evoking supernatural or spectral characters (Remaud 2022: 13, 21). The animate qualities of icebergs have long fascinated humans. Explorer Morton Moyes of the Australasian Antarctic Expedition (1911–1914), who lived alone in a hut on an ice shelf for over two months, described Antarctica as

DOI: 10.4324/9781003273080-11

'a slow-brained sentient being deceptively solid and lifeless but actually full of movement and change, with a low amoebic vitality' (Leane et al. 2020: 14–15). Similarly, French philosopher Olivier Remaud's work on icebergs blurs the line between what humans classify as alive and not alive (2022).

Today, the melting ice at the polar ice caps due to climate change, along with news reports of icebergs breaking off or 'calving' from glaciers, are usually read as markers of anthropogenic global warming. As performing climates, they are spectacular warning signs of the large-scale impact of human environmental damage. Ice, icebergs and icescapes are in many ways an ideal metaphor in a climate emergency, entities from which we are both dissociated through their incomprehensible and impressive scale, but also directly impacted in a variety of ways, such as rising sea levels and changing coastlines. This chapter explores these complex climates, their cyclical composting systems and how they might be read in contemporary performance art.

American avant-garde performance artist Laurie Anderson was attuned to the performative and animate qualities of ice in the 1970s – a time of vigorous environmental protection movements in the United States and the West more broadly. Anderson's *Duets on Ice*, first performed throughout New York in 1974 as an outdoor Kaprowian Happening and restaged several times during the last decade, sees Anderson playing a violin while laced into a pair of spotless white ice skates, the blades of the skates frozen into a pair of ice bricks. Anderson plays a 'duet with herself', live experimental electric violin over pre-recorded loops of avant-garde classical music. The music is interspersed with Anderson recounting childhood memories on topics as diverse as her late grandmother and ducks on a frozen lake (Jestrovic 2004). With the ice functioning as a miniature stage, the conceit of this performance is that Anderson plays until the ice melts – she performs at the speed of melting ice which varies according to the ambient temperature surrounding each performance.[1]

Like many land art pieces of the 1970s, Anderson's use of ice brings the more-than-human performing object into the frame to add another element of contingency to the avant-garde performance. In the early performances of *Duets on Ice*, the ice plays the role of an experimental stage. In more recent re-stagings, such as the opening of the 2013 Adelaide Festival or during the pandemic in New York in 2021, renamed *Laurie Anderson: The Weather*, the work more actively invites audiences to reflect on changing climates, melting ice, rising seas and precarious weather patterns. During these performances, an older Anderson is dressed in long white flowing robes with bleached blonde hair, her aging body made precarious by the unstable and changing ice 'stage' beneath her feet (Figure 11.1).

From a human perspective, ice and icebergs are full of contradictions. Throughout history, icebergs and icescapes have been perceived as both majestic, terrifying, sublime and highly fragile. As Elizabeth Leane writes:

> [A] berg reveals how much escapes our attempts to make meaning. In the Anthropocene, it should be no surprise that icebergs signify the paradoxes

Figure 11.1 Laurie Anderson, *Duets on Ice*, Adelaide Festival, 2013.
Photograph by Tony Lewis. Courtesy of the Adelaide Festival.

> of our times: material and symbol; threat and resource; 'pure nature' and sinister holders of hidden secrets. Icebergs are victims of humanity's destructiveness and omens of its coming doom, but also speculative sites of its ingenuity, its enterprise, its aspiration to mastery over the planet.
>
> (2022: 379)

Icebergs are shape-shifters that also function as unreliable mirrors for humanity, reflecting back what we have done to the natural world. As Remaud puts it: '[t]hey express the evolution of a world on the threshold of catastrophe' (2022: 23).

Above all, Remaud encourages us to view icescapes as more than landscapes. He describes them as living, animate, anthropomorphising them as 'landscapers' and 'sculptors who polish' (2022: 39). If humans perceive icebergs as living, he argues they will become worthy of human protection:

> If glaciers are active and reactive beings, agents and not things, essential elements in the Earth's cycles as well as partners in social life, they are undoubtably an 'endangered species' today.
>
> (Remaud 2022: 127)

Framing icebergs as a 'species' suggests their capacity as more-than-human performers in processes of dynamic composting of their hydrogen and oxygen molecules and evokes yet another kind of complex interplay of performing climates. So what would it mean to centralise the icescapes and icebergs in our reading of performances that use ice? Rather than serving as a stage as in *Duets on Ice*, what if ice was the protagonist?

Danish-Icelandic artist Olafur Eliasson is one artist who invites us to think like an iceberg by making icebergs the principal performers in large-scale installations in central sites of major European cities. In 2014, Eliasson, in collaboration with geologist Minik Rosing, first staged *Ice Watch* (2014), though the work was subsequently performed in several different locations. In *Ice Watch*, large chunks of glacial ice (around 80,000 tonnes of it) were sourced and towed from the seas around Greenland. They were transported onto land and trucked into city centres where they were displayed as 12 sculptures positioned in a clock formation. In 2015, they were placed at the Place du Panthéon, a busy Parisian roundabout in the centre of the city where cars, buses and bikes skirted the edges of their concrete stage. The icebergs were living, breathing sculptures that moved with time, melting and changing form before the spectators who were free to interact and move around them.

The documentation of *Ice Watch* shows the spectators not only observing the ice but physically engaging with it in intimate ways – embracing it, kissing it, pressing their ear to it, fingering its smooth, wet edges and craters, some even danced in slow motion beside it. Eliasson notes that he created the work in order to materialise the problem of what is ordinarily happening out of sight and mind on a warming planet:

> It is so abstract, it's so far away Greenland, it's literally out of our body and it's in our brain and I wanted simply to change that narrative of the climate from our brain and emotionalise it into our bodies.
>
> (Yalcinkaya 2018)

The embodied experience of climate change, particularly the cold ice in contact with the spectators' bare skin, offers a contrasting experience to data-driven scientific reporting on climate change. Instead, it offers a somatic experience of atmospherics, we might say a performing climates, that connects humans to their mess, to 'emotionalise' the experience of a slowly changing world in Eliasson's invented verb. Eliasson wants *Ice Watch* to be more than a tactile experience for spectators. He wants people to have an aural and olfactory experience with the artwork, to hear the cracking of the ice that releases bubbles that have been frozen within it for millennia: 'Ten thousand years ago there was 30 per cent less carbon dioxide in the atmosphere, so the smell of the ice blocks should be the smell of the air from 10,000 years ago' (Eliasson in Yalcinkaya 2018). Does this holistically embodied engagement with deep time help put the human and its relatively short blip in geological history into perspective for spectators?

The title *Ice Watch* seems to centralise the human act of watching sculptures that change before spectators' eyes, an experience that literalises Remaud's idea of icebergs as 'an open-air art gallery' of master sculptures/sculptors (2022: 11). In video documentations of the work, viewers appear to show respect, even sublime reverence as they attempt to listen and commune with the pieces of ice. Ultimately, however, it displaces the ice from its natural habitat and offers the public a very safe, palatable and digestible encounter with it. In *Ice Watch*, spectators can only view a part of the cycle of ice transformation but its deep time atmospherics are a starting point for remembering, for warding off our amnesia a little longer. These are ancient sculptures/sculptors whose formation likely predates the arrival of humans on the planet. The opportunity to watch these prehistoric treasures disappearing before our eyes and beneath our touch might be humbling.

We are reminded here of Australian artist and Muruwari, Bundjalung and Kamilaroi man Brian Martin telling Lara that Australia has the oldest cathedrals in the world in the unique rock formations on Country in outback Australia. Yet, since settlers do not see the value of these places in comparable terms, we continue to allow mining companies to blow them apart for the sake of extracting valuable minerals (see, e.g. Oliveri et al. 2022). In this revelatory statement, Martin offered a profound example of the imbrication of Country, culture, spiritual and matrilineal systems and their conjoined destruction. Eliasson attempts to foster a similar connection for Westerners by positioning the icebergs in *Ice Watch* in major Western cultural centres and cities, such as outside Bloomberg's European headquarters in London and in front of Tate Modern, though their radical displacement in such sites is deliberately marked and jarring for spectators.

The first staging of *Ice Watch* at City Hall Square in Copenhagen in 2014 coincided with the publication of the UN IPCC's Fifth Assessment Report on Climate Change. The second installation in Paris (described earlier) coincided with the 2015 UN Climate Conference COP21 and restaged to coincide with the meeting of world leaders at the COP24 climate change conference in Katowice, Poland. Yet some years before Eliasson's high profile public installations, ice was used in a performance by Tongan-Australian artist Latai Taumoepeau who was an artist-delegate for the UN climate change conference COP13 in 2007, held in Bali. At this COP, negotiations were particularly focused on Australia's responsibility to its regional neighbours in the Pacific Ocean. Australia and other Western nations have made major contributions to global warming as well as inflicting a destructive colonial presence on many Pacific Islands. These colonial occupations have been driven by capitalist-extractive interests; specifically mineral mining so extreme that it has depleted some islands to the point of being almost uninhabitable (Teaiwa 1994; Teaiwa 2015). By contrast, the Pacific Islands, which constitute 12 countries and thousands of islands of varying sizes, whose inhabitants have made little to no contribution to climate change are increasingly impacted by rising sea levels.

Taumoepeau's presence at COP13 as an Australian with Tongan heritage and family was thus significant as a witness and artist respondent to these discussions and negotiations. She found herself frustrated by the Australian government's anaemic response to the growing crisis. Taumoepeau commented: 'I was really annoyed at how Australia couldn't take responsibility for the demise of its neighbouring countries as a result of climate change and global warming' (*Body of Art* 2013). In response, she created a performance with ice, *i-Land X-isle*, a work in which the human-ice relations are vastly different from those of *Ice Watch*.

i-Land X-isle

In the first performance of *i-Land X-isle* (2012), Taumoepeau hangs tethered by ropes from a metal structure for two days. Above her suspended body, a two-tonne block of ice melts and drips onto her and the ground below. Taumoepeau wears an orange high-vis wetsuit and black wetsuit boots, suspended by white ropes that encircle her ankles, chest and neck – she resembles a tangled puppet or prisoner. Taumoepeau was bound to the ice both literally and figuratively, referencing Islander cultures which are deeply connected to oceans for food, travel between communities and spirituality. As the icy water drips onto Taumoepeau's restrained body throughout this durational work, the artist visibly and audibly expresses pain and discomfort. The slow drip of water on her brown-skinned body recalls media reports of waterboarding torture on suspected terrorists by the CIA during the Iraq War. As the ice shrinks above her, a puddle of water gradually expands on the floor beneath her. Referencing the privilege of the political leaders whose failed negotiations catalysed the work, Taumoepeau first performed *i-Land X-isle* in the harbourside setting of Circular Quay, a major tourist destination in Sydney. The framing of a vulnerable, Islander body, caught like a fish in a net, was contrasted with the harbour setting and the wealth and luxury surrounding its local and tourist spectators.

Like *Ice Watch*, *i-Land X-isle* materialises and offers spectators' proximity to ice as a metonym for the melting icecaps but Taumoepeau's Tongan body sharpens the work's critique. This is not a straightforward criticism of anthropocentrism, nor an abstracted disappearing of an important natural phenomena and its ecosystems caused by all humans. Instead, it spotlights the impact of rising sea levels on particular vulnerable communities, specifically those living in low-lying Pacific Islands. By extrapolation, the performance points the finger at those settler-colonial groups that have led the industrialisation and extractive practices which are the cause of rising sea levels and its social impacts.

Like Taumoepeau's body, the performing ice blocks in *i-Land X-isle* are firmly ensnared in the shipping ropes but, as they melt, they elide the constraints in a way that the human body cannot. The ropes in the ice suggest human attempts to master and control the movement of fresh water but, in its shape-shifting capacity, it refuses to yield to human demands. The plasticity of the two-tonne ice block is, in Remaud's terms, 'spectral' as it haunts Taumoepeau's suffering body. The dual appearance of the ice and Taumoepeau's body tethered together enhances the idea of ice as an animate and 'endangered species' (Remaud 2022). In *i-Land X-isle*, we are invited to reflect on colonial environmental mismanagement exploitation and greed and its multiple endangerments: changing oceans, lost fresh water, Antarctic habitats, Pacific Islander lives and culture.

Ice provides an ideal medium for a durational performance because, at average room temperatures, its melt gives the impression of slowing down time. Describing the ice performance *Some Things Happen All At Once* (2008–09), Carl Lavery suggests that the effect of a durational melt 'induces a feeling of loss and perhaps even a sense of helplessness, certainly a kind of fragility' (Bleeker et al. 2019: 263). In *i-Land X-isle*, the water dripping off the ice becomes a way of showing time as the kind of slow violence that human-made changes are having on Earth's environment. As Rob Nixon writes:

> [W]e also need to engage the representational, narrative, and strategic challenges posed by the relative invisibility of slow violence. Climate change, the thawing cryosphere, toxic drift, biomagnification, deforestation, the radioactive aftermaths of wars, acidifying oceans ... The long dyings – the staggered and staggeringly discounted casualities, both human and ecological ....
>
> (2011: 2)

The ecological dimensions of the ice and its relationship to Taumoepeau embody what Peter Eckersall and Eddie describe as 'slow dramaturgy' (2011). This is performance that refunctions the pace, structure and material dimensions of our time-poor, capitalist Western lives. Yet Taumoepeau's performance invites possibilities to expand the idea of slow dramaturgy, drawing in the tropical temporalities of the Pacific Islands and recalling more traditional paces of living.

i-Land X-isle is a confronting work that stages decomposition and the particular human suffering of Taumoepeau's body through an aesthetic that evokes the processes of composting. It invites durational thinking that might put human existence into perspective and, like *Ice Watch*, invites its targeted Western audiences, to see the decay of the ice in tandem with the cultural decay of Islander peoples and their rich cultural heritages. In *i-Land X-isle*, atmospheres and elements – water, sunlight, ice, air and flesh – are situated in relation to temporal and spatial fluctuations brought about by the climate emergency, its material and political dimensions.

Taumoepeau expanded her work with ice several years later in the performance *Ocean Island Mine* (2015). *Ocean Island Mine* was first performed outside Carriageworks in Sydney in 2015, in the United Kingdom in 2016 and again in Australia in 2017. In this performance, Taumoepeau silently shovels 1000 kilograms of ice from one pile to another, 40 metres apart. She wears a white, waterproof suit, white shoes and no gloves. She carries a large white spade which she occasionally drags between the two piles of ice. In contrast with the highly aestheticised block of ice used in *i-Land X-isle*, in *Ocean Island Mine*, she deploys the more familiar, accessible, everyday object of 'party ice' – the same quantity used in a bar over a day (Taumoepeau in Mangioni 2021: 45). Taumoepeau walks steadily between the deposits of ice, her facial expression remaining affectless as she performs this Sisyphean task. The scraping of the shovel on the pavers and the crunch of the ice as she digs is grating and evocative of the cold, further emphasising her Islander body's displacement from tropical climates.

The work's performance of digging references the history of extraction by colonial powers in the Pacific. In particular, it evokes the intensive phosphate mining on Ocean Island, Nauru and Christmas Island (part of the nation of Kiribati). Phosphate was a valuable fertiliser used for agriculture, predominantly in developed countries like Australia, New Zealand and Britain. Its highly lucrative extraction rendered these nations either unhabitable or hostile to human survival within decades and led to forced migrations based on changed environmental conditions. These migrations occurred in conjunction with national labour schemes in the colonial nations, where Western nations such as Australia further benefited from a cheap workforce of climate refugees (Shlomowitz and Munro 1990). In *Ocean Island Mine*, ice stood in as a metonym for the white rock of phosphate, drawing attention to colonial extractive capitalism and its exploitation of natural 'resources' and Islander people alike.

Te aba

Katerina Teaiwa is an artist, anthropologist and activist from another Pacific Island, Ocean Island, or what is traditionally known as Banaba, a small island whose closest neighbour is Nauru.[2] Teaiwa explains that all her work and life is underpinned by the philosophical idea of *Te aba* in Banaban and I-Kiribati culture, words that capture the idea that human and land (or in this case the rock that constitutes Banaba) are deeply interconnected, and that being and belonging are tied to the indissoluble nature of this bond (2015: 9). She illuminates this with a vivid explanation that when colonising forces mined and extracted phosphate from Banaba in the twentieth century, they also indiscriminately took the bones of ancestors. This included the bones of her great-great-grandmother whose remains were sprinkled all over the agricultural lands of Australia and New Zealand where they were taken up into crops and eventually consumed by the inhabitants (Teaiwa 2021) – extraction turned colonial cannibalism (and not as a metaphor!).

Te aba is part of a broader philosophy, common to most First Nations cultures, that champions human accountability towards and respect for the health and well-being of the natural world. As former Samoan Head of State, Tui

Ātua Tupua Tamasese Tupuola Tufuga Efi writes: 'The relationship between trees, animals, ocean, and mother earth is one of balance, where boundaries are respected and protected … To find solutions to our current environmental problems we must ask ourselves what it is to be a fish' (2019: xii–xiii). First Nations Oceania episto-ontologies demand then a fresh consideration of the foregrounding of the body that has been integral to Western performance art paradigms since at least the 1960s. To consider works like *i-Land X-isle* and *Ocean Island Mine* within the philosophical framework of *Te aba* is to view the performer within the relational matrix of place *and* within the natural elements where Taumoepeau situates her body in dynamic exchange. *Te aba* suggests that Taumoepeau not only represents the island, she *is* the island as the (salt) water engulfs ancestorial burial sites and food-growing land – her body shares its fate with the island. As the poet (and Katerina's sister) Teresia Teaiwa writes: 'We sweat and cry salt water, so we know that the ocean is really in our blood' (Hau'ofa 1994: 127).

Taumoepeau's body-based performances and deployment of ice as more-than-human actors encapsulate the interconnections of multiple crises. They reveal the intersections of racialised and gendered violence as well as the threat of extinction to ancient cultures that cannot be separated from discussions of habitable land, access to food and potable water. The chosen site-specific locations highlight the fact that nonwhite bodies do not have the same mobility to escape the challenges presented by a changing climate. Taumoepeau's ice can shape-shift from ominous, life-threatening water torture to banal party ice, symbolising her people and their Lands'/Rocks' subjugation, exploitation and discomfort for the benefit of colonial pleasure, freedom and wealth. The relationships between the human body and the animate aesthetics of ice in these performances are more confronting than those depicted in *Duets of Ice* or *Ice Watch*. Such performing climates reveal the geopolitical power dynamics of what ice represents in a climate emergency.

In their book *Performing Ice*, Carolyn Philpott, Elizabeth Leane and Matt Delbridge note the similarities between ice, icescapes and icebergs and performance studies as a discipline. They remark upon their shared ephemeral and continually uncanny transforming states as well as noting their significance in the Anthropocene for vividly revealing the complex interconnections of the globe, its creatures, systems and geology. They write:

> The polar regions, long thought of in the Anglophone imagination as remote, isolated places, are increasingly recognised as intimately linked, through climate, currents and sea level, to more temperate locations.
> (Leane et al. 2020: 6)

Yet Leane warns us not to expect an iceberg to yield a single meaning. We broaden this idea to ice more generally and the life forms with which it interacts as having a plasticity that makes fertile material for compost theatre (2022: 378). Artists like Eliasson and Taumoepeau are attuned to ice's capacity for decay and re-composition, shape-shifting, awe and as a mirror to human behaviour. As children going to and playing in theatres, dry ice enthralled us with its cheesy *Phantom of the Opera* stage magic. Today, in the hands of artists like Taumoepeau, the deployment of ice in performances is revealing a more radical relationality. If we can decelerate our lives enough to work at ice time, we might hear some of its ancient secrets.

Notes

1 In one performance in Genoa, Anderson chose a site next to the 'Monument to the Eternal Flame' where the heat emanating from the fire made the show half as long (Jestrovic 2004: 29).

2 Nauru is well known to Australians for being one of the islands in the Pacific where Australian governments have locked up asylum seekers who have attempted to sail to Australia.

References

Body of Art (2013). Short film, dir. Iona Reto. vimeo.com/60363446, accessed 1 June 2022.

Duets on Ice (1974). Performance Work by Laurie Anderson.

Eckersall, Peter and Eddie Paterson (2011). 'Slow Dramaturgy: Renegotiating Politics and Staging the Everyday'. *Australasian Drama Studies*. 58: 178–192.

Efi, Tui Ātua Tupua Tamasese Tupuola Tufuga (2019). 'Prelude: Climate Change and the Perspective of the Fish'. In Tony Crook and Peter Rudiak, eds, *Pacific Climate Cultures: Living Climate Change in Oceania*, Warsaw: De Gruyter, pp. ix–xiii.

Hau'ofa, Epeli (1994). 'Our Sea of Islands'. *The Contemporary Pacific*. 6(1): 147–162.

Ice Watch (2014). Performance Work by Olafur Eliasson and Minik Rosing.

i-Land X-isle (2012). Performance Work by Latai Taumoepeau.

Jestrovic, Silvija (2004). 'From the Ice Cube Stage to Simulated Reality: Place and Displacement in Laurie Anderson's Performances'. *Contemporary Theatre Review*. 14(1): 25–37.

Lavery, Carl (2019). 'How Does Theatre Think through Ecology?' In Maaike Bleeker, Adrian Kear, Joe Kelleher and Heike Roms, eds, *Thinking through Theatre and Performance*. London: Methuen.

Leane, Elizabeth (2022). 'Ice Islands of the Anthropocene: The Cultural Meaning of Antarctic Bergs'. In Russell McDougall, John C. Ryan and Pauline Reynolds, eds, *Postcolonial Literatures of Climate Change*. Leiden: Brill, 361–382.

Leane, Elizabeth, Carolyn Philpott and Matt Delbridge (2020). 'Performing Ice: Histories, Theories, Contexts'. In Carolyn Philpott, Elizabeth Leane and Matt Delbridge, eds, *Performing Ice*. Cham, Switzerland: Palgrave Macmillan, 1–26.

Mangioni, Talei Luscia (2021). 'Confronting Australian Apathy: Latai Taumoepeau and the Politics of Performance in Pacific Climate Stewardship'. *The Contemporary Pacific*. 33(1): 32–62.

McDougall, Russell, Ryan John C and Reynolds Pauline, eds (2022). *Postcolonial Literatures of Climate Change*. Leiden: Brill.

Nixon, Rob (2011). *Slow Violence and the Environmentalism of the Poor*. Cambridge, MA: Harvard University Press.

Ocean Island Mine (2015). Performance Work by Latai Taumoepeau.

Oliveri, Vicki Antonia, Glenn Porter, Chris Davies and Pamela James (2022). 'The Juukan Gorge Destruction: A Case Study in Stakeholder-Driven and Shared Values

Approach to Cultural Heritage Protection'. *Journal of Cultural Heritage Management and Sustainable Development.* https://doi.org/10.1108/JCHMSD-12-2021-0208

Philpott, Carolyn, Elizabeth Leane and Matt Delbridge, eds (2020). *Performing Ice.* Cham, Switzerland: Palgrave.

Remaud, Olivier (2022). *Thinking Like an Iceberg*, trans. Stephen Muecke. Cambridge: Polity Press.

Shlomowitz, Ralph and Doug Munro (1990). *The Ocean Island (Banaba) and Nauru Labour Trade: 1900–1940.* Bedford Park: Flinders University of Australia.

Stevens, Lara (2022). 'Sink or Swim: Performing the Iniquities of the Climate Crisis'. *Critical Stages/Scènes Critiques.* (26). www.critical-stages.org/26/sink-or-swim-performing-the-iniquities-of-the-climate-crisis.

Teaiwa, Teresia K (1994). 'bikinis and Other s/pacific n/oceans'. *The Contemporary Pacific.* 6(1): 87–109.

Teaiwa, Katerina Martina (2015). *Consuming Ocean Island: Stories of People and Phosphate from Banaba.* Bloomington and Indianapolis: Indiana University Press.

Teaiwa, Katerina Martina (2021). 'Katerina Teaiwa'. *VCA Art Forum.* 8 October. https://www.youtube.com/watch?v=fFJQC1W-qDI, accessed 12 December 2023.

Viney, Steven and Edwina Seselja (2022). 'Postcards from the Frontlines of Climate Change'. *ABC News*, 11 May. www.abc.net.au/news/2022-05-11/climate-change-effects-facts-causes-pacific-postcards/100999144, accessed 12 October 2023.

Yalcinkaya, Gunseli (2018). 'Olafur Eliasson Installs Giant Blocks of Glacial Ice across London'. *Dezeen*, 12 December. www.dezeen.com/2018/12/12/ice-watch-olafur-eliasson-installation, accessed 12 October 2023.

12 Worms

> [T]his essay aims not only to summon the elements, but also to hail agents of elemental rearrangement, to call out those composing our airs, waters, and soils. One way to tap into the elemental forces making and breaking worlds today is to pay attention to the rearrangers, the choreographers, the beings whose doings catalyze, synthesize, and stir up planetary substances and energies. We need a way to think the agencies of rearrangement. Who and what rearranges the elements on worldly scales?
>
> (Myers 2017: 123)

We begin with worms. Moving under and across the surface of the soil, hidden under decaying leaf matter and rotten vegetation, a group of red wiggler worms (*Eisenia fetida*) are changing the composition of the earth. Worldwide, there are over 6000 species of earthworms, from 20 families. In Australia, the red wigglers are an introduced European species and live alongside native Australian earthworms (Blakemore 2019). Red wigglers are prized for their ability to break down vast quantities of matter as an essential part of the composting cycle (Furlong et al. 2017).[1] Composting worms are small, only a few centimetres long and their tubular bodies are lined with tiny rings so that they feel rough to touch. Each worm-body is actually a series of segments, with muscles and external bristles (called setae) that help earthworms anchor their bodies into the earth. The worms can relax and tense each independent segment to maintain structural integrity and flexibility as they burrow.

While they are a type of earthworm, the red wigglers do not actually thrive deep underground. Instead, they are 'detritivores' who feed near the soil surface on plant litter and dung. As they feed, soil and other materials and organisms pass through the worm digestive system of the pharynx, the oesophagus, the crop, the intestine and the gizzard. The worms have glands that aid the breakdown of minerals, and they also swallow tiny stones that help their gizzard grind food into digestible amounts (Blakemore 2019). As James P. Curry and Olaf Schmidt note, worms digest a variety of materials such as 'grass and other plant leaves, roots, algal cells, earthworm setae, seeds, fungi, protozoa, fragments of arthropod cuticle, and amorphous humus' (2007: 464). Indeed,

DOI: 10.4324/9781003273080-12

the interactions between earthworms, fungi, plants and microorganisms suggest that earthworms not only improve soil fertility but also transport soil, stones and seeds, as well as producing and distributing minerals within a dynamic ecology or 'food web' (Curry and Schmidt 2007). In Natasha Myers' alternative conception of nonhuman allyship, the 'Planthroposcene', worms are active agents of rearrangement and composition, catalysing, synthesising and stirring up planetary elements (2017).

The agency of worms has long been recognised. In his final work from 1881, Charles Darwin focused on how earthworms transform dirt and matter into rich soil. Darwin found that as they eat, earthworms not only create soil, they also move it by dragging, burrowing and producing castings that rearrange and rework what David R. Montgomery calls the 'skin of the earth' (Montgomery 2012: 12). In Darwin's view:

> When we behold a wide, turf-covered expanse, we should remember that its smoothness, on which so much of its beauty depends, is mainly due to all the inequalities having been slowly levelled by worms ... The plough is one of the most ancient and valuable of man's inventions; but long before he existed the land was in fact regularly ploughed, and still continues to be thus ploughed by earthworms. It may be doubted whether there are many other animals which have played so important a part in the history of the world, as have these lowly organised creatures.
>
> (1881: np)

Darwin's observations on the importance of earthworms have been more recently extended through new understandings of soil as a dynamic system within which animals, insects, fungi, bacteria and plants intermingle (Montgomery 2012). In this way, the red wiggler worms in our compost heap reveal the interactions of multiple organisms, cycles and scales that compose and choreograph performing climates. Therefore, in beginning with worms, we are also connecting with fungi, insects, rocks, roots and death, deep time and deep listening. We see worms as a way of picturing a future for theatre that considers more-than-human timescales existing alongside human stories and recognises that making art outside of the biosphere means playing with mess and relations and learning from life-cycles and patterns that extend our connections to more-than-human atmospheres and architectures.

In 2022, dancer, photographer and urban farmer Gregory Lorenzutti was invited to work with dance and fine arts students at the Victorian College of the Arts, in Naarm/Melbourne, Australia. Drawing on family history and a lifetime of working with plants, Lorenzutti reconceived the training, rehearsal and production cycle of a dance performance as a collaboration with plant knowledge and a focus on seed growth and soil education (Lorenzutti 2022a). The initial session with the performers and designers was devoted to discussion of plant lifecycles, and the students planted seeds. As the training progressed, the seeds were watered and tended to, growing as part of the rehearsal process, and finally being featured as part of a set that was composed of living plants. The final performance, *Theatrum Botanicum* (2022), saw the performers move from external garden beds that they had constructed to a traditional black-box studio that was also populated with planters filled with soil and plants. Following the performance, the organic materials were re-potted, reused or returned to the soil as rotting organic matter and food for worms as part of the composting cycle (Lorenzutti 2022b).

However, as much as the final performance explicitly drew attention to the ways in which the human bodies of the dancers interacted with the garden beds and the botanical set, it is the process of devising this performance that challenges traditional Western approaches of teaching, making and showing. *Theatrum Botanicum* foregrounds different more-than-human temporalities, and the growth and nurturing of seeds, plants and soil, as an essential part of a biodiverse theatrical process. It also, by necessity and design, paid attention to weather systems (including air quality and levels of city pollution), seasonal differences and Indigenous Wurundjeri Woi-wurrung and Bunurong knowledges of the Country. We are particularly interested in how Lorenzutti negotiated the demands and regulations of a Western training institution – including the stipulation, as is common to health and safety guidelines for live events, that any soil used on stage be sterilised and stripped of biological agents to reduce the risk of inhalation of tiny particles and possible infection – in the creation of *Theatrum Botanicum*. It is telling that, while much of the dance performance took place in the insular and air-conditioned sphere of the black box, the development of the work and its end product throughout relied heavily on the makers'/creators'/devisors' awareness of the interconnections with more-than-human performing climates: the processes of growth and decay, shifting atmospheres and plant systems.

Despite clearly drawing on interactions between humans and the plant world, the final public performances of *Theatrum Botanicum* still revolved around the bodies of human dancers and their linear timelines and timescales. As such, the worlds of mycelium and worms are difficult to glimpse in the performance. In contrast, as Maria Puig de la Bellacasa writes on 'soil time', imagining more-than-human worlds is not simply about 'slowing down' time or celebrating nonhuman agency in isolation (2015: 214). For Puig de la Bellacasa, alternative temporalities need to also create conditions that foster material and affective ethics: 'an invitation to rearrange and rebalance relations between a diversity of coexisting temporalities that inhabit the worlds of soil and other interdependent ecologies' (2015: 215). In Puig de la Bellacasa's terms, we are drawn to practices of 'intensification' and 'involvement' rather than technoscientific innovation and production (2015: 197). Returning to worms and the cycles of compost, we suggest that theatrical models of slow and deep dramaturgy are less about teleology and reducing the pace of linear timelines and more about revealing the everyday intermingling of us as practitioners, educators, spectators and scholars in affective intensities found across a multiplicity of timescales.

In her book chapter 'Squish, squelch, shlshlshlurpppp', artist, scholar and performance maker Norie Neumark reflects on the ways in which human listeners were affected by the more-than-human voices of rotting compost and worm movement in Out-of-Sync's installations, *Waiting* (2017) and *It's Still Life: Listening to Compost* (2019). Out-of-Sync refers to Neumark's collaboration with another Melbourne-based artist, Maria Miranda, and their durational practice of caring, listening and attempting to create 'mutual attunement' between human and more-than-human animals (Neumark 2022: 225). Building on their existing everyday recycling and composting practices, Neumark and Miranda began feeding worms, making compost and documenting their gardening as a form of art. They later placed microphones into a worm café and compost pile, being careful not to disturb the worms or disrupt their environment. They began listening to the sounds of worms and recording the sounds of compost being made.

The initial performance-installation, *Waiting*, was featured as part of the MoreArt Festival in 2017 at an urban train station in the suburb of Coburg. Out-of-Sync uses the Moreland (now Merri-Bek) Station waiting room, a space enclosed on three sides with seating for passengers to wait for their train. They populate the walls with close-up images of compost worms situated above a display of potted plants, a bucket of worm castings and a rudimentary screening and speaker system on which they play videos and their recordings of worm sounds (Figure 12.1).

In the station waiting room, usually a mundane place of transitory moments and meetings, a video projection shows Neumark, wearing glasses and headphones, standing in front of the red-brick wall and windows typical of a 1950s Australian suburban house. Birdsong can be heard in the background. Neumark listens intently, slowly swaying and with her eyes scrunched closed in concentration. She begins to make a series of slurping, sucking and clicking noises. Later in the video, Miranda (also wearing headphones) replaces Neumark and commences softly clicking, chirping and slurping. It is evident that, as they listen, Neumark and Miranda are trying to echo and interact with the 'wormy compost voices' they have earlier recorded. As Tessa Laird writes:

> The strange facial tics and contortions of Out-of-Sync as they become–worm are almost alien in their embrace of other-worldliness (one of the commuters exclaimed to the duo, "You speak worm!").
>
> (2018: 64)

As they transform the station waiting room into an exhibition and education space, Out-of-Sync attempts to respond to the affective dimensions of the worm-song and invite others to do the same through witnessing their

Figure 12.1 Out-Of-Sync, *Waiting*, 2017. Photograph by Maria Miranda. Courtesy Maria Miranda and Norie Neumark.

sonic attunement (Neumark 2022; see also Chapter 4). At each peak hour, as passengers streamed from the arriving trains, Neumark and Miranda would engage in dialogue with interested passers-by or offer commuters the waiting room as a place to sit with worms.

Indeed, one way of considering this attunement is as an extension of Headphone Theatre practice, commonly linked to explorations of human difference as personified by the unique qualities of an individual voiceprint (Paterson 2010). However, where the headphone practices of artists such as Anna Deavere Smith or Roslyn Oades frequently reveal tensions between class, age, gender and ethnicity (Paterson 2015), Neumark and Miranda take seriously their affective relations with the more-than-human scales, sounds and times that exist around us – perhaps creating a form of ethical creativity or what Isabelle Stengers would call 'reciprocal capture' (2010).

In attempting to capture and share the actions of composting worms, *Waiting* from Out-of-Sync offers a playful illustration of the ways that art and performance might seek to stage strange and messy interminglings of matter and being. Alongside this artistic and philosophical engagement, a personal connection to ecological entanglement appears to shape such practice. Neumark and Miranda openly admit to the 'affective lure' of the worm voice and the ways in which their worm performances are an ethical provocation to themselves, to try and avoid extractive practices and acknowledge worm intelligence and creativity (Neumark 2022).

The mutual connection and acknowledgement of more-than-human agency that we find in Out-of-Sync's process is rarely part of contemporary biospherical Western theatre and performance. Yet, longstanding concepts of dynamic interconnection are commonplace to many First Nations' understandings of kinship. In a recent reflection on the devastating loss of life in the Black Summer bushfires of 2020 that burnt approximately 30,000 square kilometres of Australia, Waanyi author Alexis Wright offers a eulogy to honour and express sorrow for the more-than-human lives lost. In Wright's words:

> In the Aboriginal world, we deeply understand that all animals that belong to this land are our close relatives. We understand the importance of our close inter-connectivity with everything in our world. This is the ancient wisdom that we must follow. If we care for our country, the country will care for us.
>
> (2023)

Wright honours the three billion living creatures killed in the fires, as well as the '240 trillion insects, beetles, butterflies, moths, spiders, yabbies and other invertebrates' whose ancestral homelands were lost in the fires (2023). Wright's eulogy draws attention to human culpability in the disaster and suggests that people should seek forgiveness and 'speak honestly' to address future climatic conditions across the planet (2023).

We find that Wright's eulogy is also an invitation to artists and scholars. It is a statement that underscores, again, that business-as-usual will not address an impending climate disaster. It is an invitation to change. For us, this means rethinking the ways we make, consume, celebrate and speak about theatre and performance. We do not need to look far to see the connected legacies of colonialism and changing climates. As Lara and Denise Varney put it, we must instead find examples of making art that 'can rehearse nonextractivist ways of engaging with humans and nonhumans that align with and draw from Indigenous knowledge around kin and care, a philosophy of living with and not on the land' (Varney and Stevens 2023: 60). We must find ways of moving outside of the biosphere of insular Western practices and thinking, to reanimating theatre's status as a *uniquely compostable artform* capable of interacting with beings, scales, atmospheres, lifecycles and times beyond human perception. When we think of worms, we also think of bacteria, fungi, protozoa, nematodes, arthropods and gastropods, we think of rot and regeneration. We think of castings! (Said the actor to the worm: 'Aren't castings a load of s**t'). We think of theatre and performance that acknowledges its capacity for interconnection. We think of theatre that reveals human and more-than-human relations *as part of the composting connections* in performing climates.

Note

1 Composting earthworms like red wigglers are key to large-scale 'vermicomposting' schemes, which are argued to be a sustainable solution for processing the 598 million tonnes of municipal organic waste and 800 million tonnes of human-produced organic waste produced globally each year (Furlong et al. 2017).

References

Blakemore, Rob (2019). 'Australian Earthworms: Factsheet'. Australian Museum, 25 February. www.australian.museum/learn/animals/worms/australian-earthworms, accessed 11 December 2023.

Curry, James P and Olaf Schmidt (2007). 'The Feeding Ecology of Earthworms: A review'. *Pedobiologia*. 50(2007): 463–477.

Darwin, Charles (1881). *The Formation of Vegetable Mould, Through the Action of Worms, With Observations on Their Habits*. London: John Murray.

Furlong, Claire, Nalika S Rahapaksha, Kevin R Butt and Walter T Gibson (2017). 'Is Composting Worm Availability the Main Barrier to Large-Scale Adoption of Worm-Based Organic Waste Processing Technologies?'. *Journal of Cleaner Production*. 164: 1026–1033.

It's Still Life: Listening to Compost (2019). Art installation, Out-of-Sync.

Laird, Tessa (2018). 'In the Beginning There Was the Worm: Animal Voice Beyond the verbal'. *Artlink*. 38(1): 58–65.

Lorenzutti, Gregory (2022a). 'Director's Note'. Credits for *Theatrum Botanicum*. https://finearts-music.unimelb.edu.au/showcase/theatrum-botanicum/theatrum-botanicum-credits, accessed 2 October 2023.

Lorenzutti, Gregory (2022b). 'Microorganism and Macro-movement: Soil & Food Cultivation for a Dance Ecology'. Presentation for Performance Studies international conference #27. www.gregorylorenzutti.com/germinations/blog-post-title-one-86mrb-z4fwc-s6xfb-shh75-5yk6n-3x9ya-4s26l-x8sr9-p85zt-2k55t-kc2r5-gzw9w, accessed 4 September 2023.

Montgomery, David R (2012). *Dirt: The Erosion of Civilizations*. 2nd edn. Berkeley: University of California Press.

Myers, Natasha (2017). 'Photosynthetic Mattering: Rooting into the Planthroposcene'. In Line Marie Thorsen, ed., *Moving Plants*. Næstved, Denmark: Rønnebæksholm, pp. 123–130.

Neumark, Norie (2022). 'Squish, Squelch, shlshlshlurpppp'. In Sharon Jane Mee and Luke Robinson, eds, *Sound Affects: A User's Guide*. New York: Bloomsbury Academic, pp. 224–236.

Paterson, Eddie (2010). 'Paperless and Penless: Headphone performance, audioscripting and new approaches to writing for performance'. In Catherine Cole, Marcelle Freiman and Donna Lee Brien, eds, *Strange Bedfellows: Refereed Conference Papers of the 15th Annual Australasian Association of Writing Programs Conference*, pp. 1–9. www.aawp.org.au/publications/the-strange-bedfellows-or-perfect-partners-papers

Paterson, Eddie (2015). *The Contemporary American Monologue: Performance and Politics*. London and New York: Bloomsbury.

Puig de la Bellacasa, Maria (2015). 'Making Time for Soil: Technoscientific Futurity and the Pace of Care'. *Social Studies of Science*. 45(5): 691–716.

Stengers, Isabelle (2010). *Cosmopolitics 1*. Minneapolis: University of Minnesota Press.

Theatrum Botanicum (2022). Performance work by Gregory Lorenzutti.

Varney, Denise and Lara Stevens (2023). 'Australian Coal Theatrics'. *TDR: The Drama Review*. 67(1): 57–61.

Waiting (2017). Art installation, Out-of-Sync.

Wright, Alexis (2023). 'The black summer bushfires killed 3 billion animals. They are our relatives; they deserve to be mourned'. *The Guardian*, 1 April. www.theguardian.com/australia-news/2023/apr/01/alexis-wright-black-summer-bushfires-vigil-sydney-speech-3-billion-animals-killed-are-our-relatives-they-deserve-to-be-mourned, accessed 12 June 2023.

13 Unicorn

Let us begin with a mythical creature – a one-horned beast, a magical pony, a rainbow-pooping heroine, a transcendent being. The earliest descriptions of what is now known as the 'unicorn' are commonly traced to the writings of Ctesias of Cnidus in around 398 BCE, who described an Indian wild ass with a single horn (Lavers 2009: 1). In describing this animal, Ctesias – a physician based in Ancient Greece and Persia – laid the foundations of a 2500-year-old myth that brings together Eastern and Western understandings of the more-than-human world. In the words of Chris Lavers, Ctesias' writings offer 'a rich and strange mish-mash of Eastern geography, zoology, botany, medicine, anthropology and nonsense' (2009: 4–5). It is no surprise, perhaps, that the image of the unicorn has continued to confound categorisation. The unicorn is a hybrid. It is real and not-real at the very same time. A theatrical paradox.

For her third birthday party, our daughter insists on a unicorn-and-rainbow theme: costumes, cake, pin the tail on the unicorn. The mermaid Ariel is briefly floated as a possibility, along with that inescapable Ice Queen Elsa, who seems to have more of a problem with global cooling than global warming, but ultimately the unicorn wins out. It is a sign of our daughter's recent attempts to aggressively assert herself as a girl and the cultural symbols that she sees modelled as associated with that gender: pink, princesses, fairies. But it is also a sign of her white privilege, her unrepentant optimism, her belief in magic before a time when she will come to understand that she is living in a climate emergency. Before a time when she knows that the narwhal, a whale with a long, spiralled tusk, which is often described as the unicorn of the ocean, is endangered due to climate change. Before a time when she understands that the 'Siberian unicorn' *Elasmotherium sibericum* – an enormous one-horned rhinoceros that once resided in Eurasia alongside humans – is now extinct due to anthropogenic causes.

Unicorn Capitalism

Scarily enough though, our three-year-old is not alone in her belief in unicorns. So prevalent was climate denial among the ranks of elected representatives in the US during the Obama administration that in 2013, Obama's Climate Campaign

DOI: 10.4324/9781003273080-13

Team Manager, Ivan Frishberg, evoked the creature in an attempt to ridicule those who were ignoring the science of climate change. Frishberg famously said: 'The sticky thing about the truth is that it's the truth whether Congress likes it or not. Unicorns don't exist, climate change is real' (Houlton 2013). A widespread refusal to believe scientists about global warming has persisted in the US and elsewhere. President Trump was once described by his former advisor Sean Spicer as 'a unicorn, riding a unicorn over a rainbow' for his capacity to turn a seemingly disastrous political comment into a spectacularly successful assault on his opponent. Trump's rainbow jumping is evident in his ability to experience few repercussions for violent, misogynist and racist behaviour and remarks, as well as his flagrant and performative denial of climate change. Trump has variously described climate change as 'mythical', 'non-existent', 'an expensive hoax' and, in 2012, claimed that it was 'created by and for the Chinese in order to make US manufacturing non-competitive', skilfully combining denial with racism (Cheung 2020). Trump subsequently claimed that this remark was a joke.

But Frishberg was wrong. Unicorns did and do exist, and that is part of the problem. In the same year as Frishberg's comment, venture capitalist Aileen Lee coined a new term for privately owned tech startup companies that became worth more than $US 1 billion: 'unicorns' (Lee 2013). The term celebrated the rarity of such companies and added some shimmer and sparkle to capitalism and its markets, whose mystical aura had been dulled by the 2009 Global Financial Crisis. Lee's term was adopted in the American reality TV show *Unicorn Hunters*, launched in 2021. The premise of the show is that its hosts seek out the next generation of tech company 'unicorns' and encourage the judges and viewers to invest in them. Today, the 'unicorn capital' of the world has moved from California to Beijing. Attempting to wrestle back this title, a significant portion of Silicon Valley startups are now focusing on building technologies that will enhance the implementation, design, availability and use of green technologies, while also generating a truckload of profit. Such a phenomenon further illustrates what Naomi Klein calls 'disaster capitalism', based on Milton Freidman's neoliberal economic principles and his idea that 'only a crisis – actual or perceived – produces real change' (Friedman in Klein 2007: 6). Klein calls it 'the intersection between superprofits and megadisasters' (2007: 9): an intersection at which the unicorn fantasy is alive and well.

For many decades, climate denialists have spruiked the hope of the techno-fix for global warming. Ignoring the fact that man-made technologies are the cause of climate change, they argue that new technologies will save humans and the planet. The best-known techno-fixes are geoengineering propositions of carbon dioxide removal and 'solar radiation management' (Nicholson 2020). 'Removal' here refers to capturing carbon from the atmosphere and storing or 'sequestering' it underground (which should be more accurately described as re-location rather than removal). 'Solar radiation management' includes options such as further polluting the atmosphere with aerosols such as sulphur in order to dim the sun's rays and thus reduce planetary heating. These ideas are most strongly advocated by Western nations and corporations with the largest carbon footprints because they justify the approach of business and growth as usual. Such technofixes take man's rationalism to its most irrational, logical end-point. They represent the cyborgian myth of hope for humanity's survival, the unicorn in the climate emergency.

A telling image of the cyborg-unicorn appears in the online and mobile videogame *Robot Unicorn Attack* (2010), from Spiritonin Media Games. This queer parody of a techno-future features an addictive dance soundtrack and a robot unicorn firing rainbows from its bum. The unicorn is in constant motion, running across the screen while the player tries to collect crystals of power. The pace of the game is frantic – the unicorn cannot stop – it is always on the move, galloping forward. The landscapes of the game are barren, as if the denuded rockfaces and hyper-coloured sky are a portent of imminent environmental collapse. If we look beyond the foreground of neon-rendered nostalgia and perpetual progression, this videogame is set amid a world in which climate disaster is clearly a reality.

Horny Creatures

Robot Unicorn Attack leads us to consider a more complicated rendering of the unicorn found in the 2016 performance *Unicorn, Gratitude, Mystery* (2016) by American feminist performance artist Karen Finley. Riffing on the proliferation of unicorns and their proximity to wealth and power, the show begins with Finley bursting onto stage dressed as a unicorn, shimmering white with a tussled, blonde mane. Throughout the show, she performs parodies of Hillary Clinton and Donald Trump, the blonde mane conveniently transferrable between these icons of power. Finley begins an earlier version of the work, a published monologue titled *Unicorn, Gratitude* by setting the scene with what she describes as the 'Aftermath':

> When my liberal mood is upset
> I just can't take the news, the media, climate change
> Like a day when there is a tragedy
> or something upsetting
> Your candidate's email is being hacked
> Or the election is rigged
> Or your pussy is grabbed
> Or that a wall is being built
> Or the farmer's market is out of frisée!
> Endive will not do
> Radicchio will not do
> Sorrel will not do
> Fiddleheads will not do.
> What do you mean the farmer's market is out of frisée and champagne vinegar?
> What do you mean freeze dried mango is out of stock?
> The world is falling apart and I was planning on having frisée this weekend!
> EVERYTHING GREEN IS AGAINST ME
>
> (2018: 154)

Climate change here appears as an inconvenient truth for the privileged, white subject, on a par with not being able to find the right kind of lettuce for the weekend meals. Her lament that 'EVERYTHING GREEN IS AGAINST ME' recalls lines from Caryl Churchill's climate plays where the natural world asserts its agency and self-weaponises to fight the human populations already at war: 'the weather here's on the side of the Japanese' (Churchill 2000: 43) or 'Swimming pools engulfed the leisure centres and coffee ran down the table legs' (Churchill 2016: 12).

From this anxious, liberal subject position, who sees the end of the world in no freeze-dried mango stock, Finley invites audiences to imagine a unicorn:

> Who doesn't love a Unicorn?
> I love a Unicorn on a backpack or a fanny pack
> It reveals mystical mythical enchantment
> The white-horned creature alone
> With a rainbow world to adore
> The white Unicorn encircles my throat as a choker
> With iron-on glitter for an all-cotton t-shirt
> It is the illusion, the magical kingdom within
> Of enchanted whiteness and purity without
> Imagined mythical extinct being in pure white
> Neoliberal creature
>
> (2018: 156)

Finley's Unicorn (which she spells with a capital U and makes clear this is You, you readers, you sitting in the audience) is ostentatiously white, privileged and entitled. It is not a rainbow-blasting saviour but instead, a lazy, white, yuppie (Figure 13.1).

Finley starts by mocking the privilege of the Unicorn and the fantasy it promises of a perfect, clean, carefree world. Finley's Unicorn is conscientious about sustainability, woke on gender, race, climate and its consumer politics, a rhetoric it uses for virtue signalling. This façade hides its neoliberal core where it clutches desperately to its bourgeois lifestyle:

> The Unicorn does not like Styrofoam under any circumstances…
> The Unicorn likes its kale massaged *please*
> The Unicorn likes its horn massaged with almond oil *please*
>
> (158–159, italics in original)

We can assume the horn is a phallic symbol in Finley's psychoanalytic piss-take. We can assume the unicorn eats carbon and shits 'clean' coal, a notion that we have heard regularly spruiked by the climate-denying major parties in the US and Australia's climate wars between the years 2000–2021. It is also a theme that we come across in our son's *Thomas the Tank Engine*

Figure 13.1 Karen Finley, *Unicorn Gratitude Mystery*, 2016.
Photo: Dona Ann McAdams.

book collection with its celebration of coal-fired steam trains. Indoctrination starts young.

Not long after her ironic setup of the Unicorn's virtues, Finley turns on her figurehead. She begins accusing the Unicorn of promising rainbow realities but never magically appearing at the moment when it is most needed. In the recording of the performance, her voice gets increasingly agitated and crescendos:

> Was the Unicorn there for Sandra Bland?...
> Where were you, Unicorn, for those in the church in Charleston?
> Where were you, Unicorn, as Trayvon opened the rainbow skittle?
> Where were you, horned creature, when Eric Garner gasped for air?...
> Take your white lies and your medieval terror liberal enchantment bullshit

Hell to you, you white saviour beast
Damn your white preciousness and impossible chivalry that
I am part of

(2018: 161)

At a time of growing climate catastrophes and when black Americans are being murdered in broad daylight throughout the United States by white law enforcement officers, those who believe the ideology of white superiority – the supposed heroes, the Unicorns – evaporate in a cloud of their own emissions. Finley concludes the monologue by saying: 'There is no Unicorn for us/Three white-horn rhinos are all that are left' (2018: 161–162). The last line reminds us that we are living in the sixth mass extinction, this one caused by Lee's unicorns, highlighting the entanglement of the climate emergency with white privilege, neoliberal capitalism and narratives of denial by those in positions of power.

Finley's satirical depiction of a unicorn that represents white privilege on a dying planet is echoed in a 2021 report issued by Friends of The Earth. The report, titled *Chasing Carbon Unicorns: The Deception of Carbon Markets and "Net Zero"*, focuses on the intersectional inequalities at the core of the climate crisis. It points out that the biggest polluting nations should be the first to stop digging up and burning fossil fuels and recognise their impact on the lesser polluting nations, particularly those in the Global South and Indigenous communities. Above all, the report critiques the pledges of major polluting nations to 'net zero' as a dangerous promise that conceals the plans of governments and big business to surreptitiously grab large areas of land and resources from Indigenous Peoples, peasants and local communities in order to meet the carbon offsets that the net-zero pledges require. Its authors argue that these pledges are a 'smokescreen' that conceals the broader goal of continuing the fossil fuel business as usual, while relying on further colonising of massive areas of lands to offset emissions (Stabinsky, Bhatnagar and Shaw 2021: 4).

As the report's authors describe it: 'Fossil futures require carbon unicorns … fanciful imaginings of how we might solve the climate crisis without needing to eliminate the burning of fossil fuels' (2021: 14, 30). They are not only critical of using untested geoengineering technologies on the basis of their dubious scientific feasibility but also of the ways in which they ignore Indigenous modes of living with the Land as spiritually and materially animate places of belonging. Instead of rainbows and sparkles, their performative logo is an inky, black unicorn that differs starkly from the unicorn of the popular imagination or Finley's white Unicorn. The carbon unicorn leaves a messy, polluting trail behind it, with a tail attached to a deathly, black dollar sign.

The Friends of the Earth report takes its title from research done by human geographer Wim Carton. He describes negative emissions as a 'convenient fiction' and uses the language of performance to understand how these ideas are marketed to the general public as plausible possibilities for avoiding climate catastrophe (2020: 36). Carton is critical of the way in which the IPCC reports integrate mainstream economic theories (based on growth) into their future modelling for coping with the effects of global warming without challenging existing neoliberal capitalist frameworks. He writes that:

> by [negative emissions] being included in the authoritative assessments of the IPCC, such pathways do not just describe possible climate futures but potentially help bring them into being—that is, they *perform* certain futures as seemingly legitimate, necessary, and desirable
> (2020: 39, italics in original)

Here the idea of IPCC reports that 'perform' points to the ways in which language can normalise or validate ideas, such as negative emissions, without questioning their legitimacy. Carton argues that reports such as the IPCC are 'profoundly ideological' and are being used to justify business-as-usual (2020: 36). This is the kind of understanding of technological and institutional performance that Jon McKenzie theorised in *Perform or Else* (2001). These imaginings, or we might say stagings, of different possible futures help actualise the very outcomes they seek to avoid.

Sparkle Princess

It is impossible to discuss unicorns, carbon or otherwise, without also mentioning glitter, that shiny dust that makes them sparkle. In recent years, glitter has been an object of study with regards to human pollution as well as being employed as a philosophical concept for grappling with the Anthropocene (Rose 2017). In the past decade, glitter, as one form of microplastic, has been criticised as a dangerous pollutant of the oceans that ends up entering more-than-human and human food chains. There have been calls to have glitter banned for its polluting effects. The fashion industry and cosmetic companies have attempted to rebrand glitter by making ecoglitter or 'green' glitter, which is biodegradable. These glitters have their own problems, including causing eutrophication: the unnatural addition of nutrients to freshwater and marine habitats. Behind these multimillion-dollar industries, the qualities which make the glitter shimmer – aluminium or mica – are toxic substances that cause substantial environmental damage through their mining. Aluminium is now rare and thus an increasingly valuable commodity. Mica has been linked to the largest illegal mining operation in India, which relies on a large workforce of children (Bramley 2018). Glitter, we might say, has become a metaphor for how we in the West dress ourselves up to perform glamour and wealth for one another – an artificial sparkle of celebrity that, like much of the fashion industry, obscures the labour and exploitation of people and nature that underpins it.

Anthropologist Deborah Bird Rose, who lived and worked closely with Australian Aboriginal communities for over 30 years, explains the concept of 'shimmer' as conceptualised by Aboriginal people in the Victoria River region of Australia's Northern Territory (2017). The shimmer, Rose elucidates, illustrates the dynamics of what she describes as our 'multispecies worlds' (2017: 53). The idea comes from the Yolngu word *bir'yun*, which describes both aspects of artistic practice such as ritual, dance and painting and the world in motion, a vibrant, vibrating world that is alive. In artistic practice, it refers to capturing such movement and drawing the eye. More than this, 'shimmer' is a 'capacity to see and experience ancestral power' (2017: 55). Shimmer captures the transgenerational interconnectivity of our world and its creatures, the everywhen.

The shimmer's figuring of a past and future vibrancy of living and non-living matter stands in contrast to Western linear temporality that abruptly divides living from dead. Rose argues that the Indigenous philosophical idea of shimmer makes our species more aware of our responsibility and long-term impact in the world. As she puts it:

> extinction cascades drag shimmer from the world ... With it comes a burden: the commitment to bear witness to the shimmering, lively, powerful, interactive worlds that ride the waves of ancestral power ... Shimmer comes with the new growth, the everything-coming-new process of shininess and health, and the new generations
>
> (2017: 53, 54)

Rose uses poetic images to illuminate this Indigenous concept of a world that radiates animacy and marks time. Let us not forget that Finley too is famous for her use of glitter. In her performance *The Constant State of Desire* (1986), she slathers her body in crushed raw eggs before coating herself in gold glitter. The performance is a commentary on the patriarchal expectation that women will still dress up glamorously after being degraded or abused, to maintain the illusion of their bodies as pristine objects of beauty for consumption by men. Glitter, unicorns, desire, consumption, inequality, exploitation, vulnerability and greed are themes spanning her oeuvre, never more relevant than in times of ecological crisis.

Finley's *Unicorn, Gratitude, Mystery* is not explicitly a performance about environmental crisis, and yet its titular figure embodies the contradictions that have slowed meaningful climate action. Finley demonstrates how white privilege and the chimaera of the sparkling Unicorn are today symbols of economic, social and political clout. Meanwhile, the 'real' unicorns of the business world are the most powerful and visible figures in a global free market economy that still circles around Western narratives of progress, growth and development. The ongoing faith in saviour technologies, damage to oceanic and terrestrial food chains, the exploitation of women and children and their exposure to toxic substances and the colonial demeaning of First Nations knowledges are all entangled in the contemporary unicorn narrative. Horny techno-capitalist unicorns exemplify cultures of extraction, treating nature and others as disposable commodities. Further, they are tied to a belief in a capitalist technofix that enables unicorns to thrive and that slows down our capacity to respond to the climate emergency. As Western nations increasingly chase carbon unicorns, most of the human species is losing sight of the shimmer in the natural world. Instead, we in the West largely remain narcissistic three-year-olds, entranced by our own refracted reflection in the mirrorball of our fantasy of mastery.

References

Bramley, Ellie Violet (2018). 'Losing Its Sparkle: The Dark Side of glitter'. *The Guardian*, 22 January. www.theguardian.com/fashion/2018/jan/21/losing-its-sparkle-the-dark-side-of-glitter, accessed 28 November 2023.

Carton, Wim (2020). 'Carbon Unicorns and Fossil Futures: Whose Emission Reduction Pathways Is the IPCC Performing?' In J.P. Sapinski, Holly Jean Buck and Andreas Malm, eds, *Has It Come to This? The Pitfalls and Promises of Geoengineering on the Brink*. New Brunswick, NJ: Rutgers University Press, pp. 34–49.

Cheung, Helier (2020). 'What Does Trump Actually Believe on Climate Change?' *BBC News*, 23 January. www.bbc.com/news/world-us-canada-51213003, accessed 19 October 2023.

Churchill, Caryl (2000). *Far Away*. London: Nick Hern Books.

Churchill, Caryl (2016). *Escaped Alone*. London: Nick Hern Books.

Finley, Karen (2018). *Grabbing Pussy*. New York: OR Books.

Houlton, Sarah (2013). 'Unicorns don't exist'. *Chemistry & Industry*. 77(8): 23.

Klein, Naomi (2007). *The Shock Doctrine: The Rise of Disaster Capitalism*. London: Allen Lane.

Lavers, Chris (2009). *The Natural History of Unicorns*. New York: William Morrow.

Lee, Aileen (2013). 'Welcome To The Unicorn Club: Learning From Billion-Dollar Startups'. *TechCrunch*, 3 November. www.techcrunch.com/2013/11/02/welcome-to-the-unicorn-club/, accessed 19 October 2023.

McKenzie, Jon (2001). *Perform or Else: From Discipline to Performance*. New York: Routledge.

Nicholson, Simon (2020). 'Solar Radiation Management'. *Wilson Centre*, 30 September. www.wilsoncenter.org/article/solar-radiation-management, accessed 12 December 2023.

Robot Unicorn Attack (2010). Computer Game, Spiritonin Media Games.

Rose, Deborah Bird (2017). 'Shimmer: When All You Love Is Being Trashed'. In Anna Tsing, Heather Swanson, Elaine Gan and Nils Bubandt, eds, *Arts of Living on a Damaged Planet: Ghosts and Monsters of the Anthropocene*. Minneapolis: University of Minnesota Press, pp. 51–63.

Stabinsky, Doreen, Dipti Bhatnagar and Sara Shaw (2021). *Chasing Carbon Unicorns: The Deception of Carbon Markets and "Net Zero"*. Amsterdam: Friends of the Earth International.

The Constant State of Desire (1986). Performance work by Karen Finley.

Unicorn, Gratitude, Mystery (2016). Performance work by Karen Finley.

14 Tree

with Angie Abdilla

> When you create something, eventually someone is going to look at it and talk to you about it, or if you make things together you collaborate. Those conversations and relationships have been going on for millennia in our culture. It is about connection and it's about relationality to people, to entities, to Country. It's always about some sort of relationality to the world, that might be political or involve a reclamation of culture, or ancestral memory, or – whatever it may be – there's a deeper connectedness
>
> (Martin 2022)

We begin with roots. Metres underground, an anchoring root of the *Agonis flexuosa* plunges downwards, working its way through and around rocks, the thimble-sized root-tip forcing a path through tightly packed sediment, seeking cracks and following the scent of deep water. The primary root from the *Agonis* and the taproots that branch from it create a series of anchor points that keep the tree stable in high winds and drought. Closer to the surface, 30 centimetres into the soil, smaller fibrous roots emerge from the stem and sides of the taproots. At Wooragee, where a natural spring lies at the top corner of the farm, water flows over and under sheets of granite beneath the earth's surface, and the *Agonis* is aware of this. It sends these quick-growing root-forms to seek water and nutrients throughout its critical root zone, which mirrors the dimensions of the leaf canopy above. The fibrous roots filter liquids through the root-skin (*epidermis*) and transport liquids through xylem tissue, which carries water and dissolved minerals from the soil to the leaves, and phloem tissue, which carries sugar and nutrients produced in the leaves by photosynthesis to other parts of the tree. Within each fine root, masses of tiny cells are rapidly proliferating, decaying and regenerating.

The anchoring and feeding roots of the *Agonis* are not alone under the earth. As well as transporting water and energy, roots live in the soil among insects, worms, spiders and interact with fungi, algae and protozoa. Mycorrhizae colonise the tips of the fine feeder roots – helping them absorb nutrients and connecting the tree to fungal communications – and the soil ecosystem breaks down organic material that the *Agonis* uses as food. When we begin with tree roots, we picture a continuous collaborative compost

DOI: 10.4324/9781003273080-14

choreography. This choreography stretches beyond a human lifespan, and involves elements that are both microscopic: bacteria, protozoa, nematodes, fungi networks; and macro-climatic: atmospheres, weather systems, celestial movements.

In the 2022 immersive installation *Ngapulara Ngarngarnyi Wirra (Our Family Tree)*,[1] Indigenous data sovereignty principles, the cultural programming of machine learning, and spatial technologies were integrated within a unique approach to video and sound art. The artwork is the result of stewarding the repatriation of a culturally significant data set, and the investigation of Indigenous spatial awareness, driven by 'Technology as cultural practice' (Harle et al. 2019) – deep-time ways of knowing that are transdisciplinary in nature and, bound within a cultural sensibility. The result was a liminal space representing Country and kin, which created an opportunity for people to embody the metaphysics of connection and relationality.

The project was a collaboration between former Australian Football League (AFL) athlete, Adnyamathanha man Adam Goodes, Baden Pailthorpe and Professor Angie Abdilla, palawa artist, filmmaker, designer and founder of Old Ways, New.[2] As Brian Martin notes in the epigraph to this chapter, creating art can lead to exciting conversations as well as potential collaborations, and we were fortunate to connect with Angie Abdilla as we were writing about *Ngapulara Ngarngarnyi Wirra*. As our co-author for this chapter, Abdilla situates her own expertise and experience making the work alongside our responses as spectators. Abdilla led the creative process, stemming from the Indigenous Protocols for Artificial Intelligence (IP//AI) body of research. IP//AI incorporates ways of knowing, relating, approaches to knowledges and methods of practice for automated technology systems, through an Indigenous standpoint and a cultural paradigm (Abdilla et al. 2021).

During the six-year development of the artwork, the creative team travelled to Adnyamathanha Country, where local Elders shared with them an ancient river red gum known as Wirra. The Wirra became a central representation of the artwork: Kin; a representation that was paired with the other central representation: Country. Inspired by Indigenous relationality, spatiality, and a commitment to situating the work beyond a creative practice, Abdilla identified the Adnyamathanha kinship system as the foundation for the cultural practice. The making of the artwork, including all creative and technological decisions, was then appropriately guided by the logic of the kinship system.

An Indigenous kinship system works to moderate human genetics through marriage rules and blood types to maintain health and avoid birth defects within community. At the centre of Adnyamathanaha (translating to 'people of the rock') kinship system are two key moieties, Arraru (warm/north wind) and Mathari (cool/south wind). All humans belong to one of these two moieties, which signifies who you are both cared for and cared by and the types of relationships that are good or taboo within the relational system. The kinship

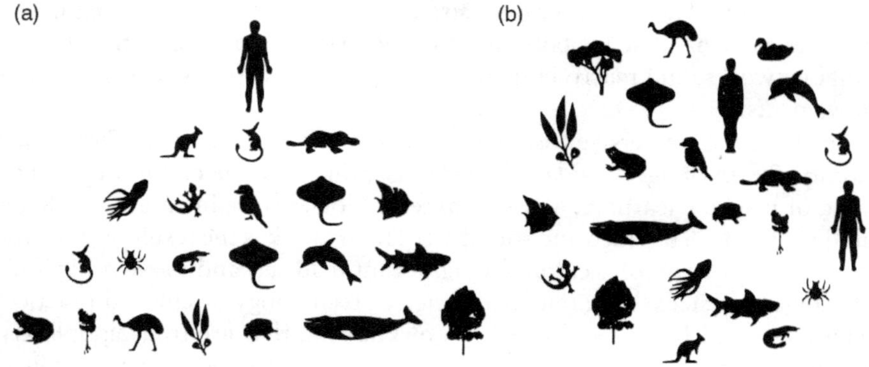

Figure 14.1 Country-Centred Design diagrams. Left-hand figure: Western worldview and evolution of systems design. Right-hand figure: Indigenous worldview and evolution systems design.

Abdilla et al. (2021: 6)

system then expands out to totems, connecting the system of care to animals and, in some instances, plants.

Within Indigenous ways of knowing, 'the rock has value, meaning, and place, as do human beings and the animal, plant, cosmological and metaphysical worlds combined' (Abdilla and Fitch 2017: 8) (Figure 14.1). As Abdilla expands:

> IKS [Indigenous Knowledge Systems] can nudge the existential compartmentalism of Western techno-science into another realm of interrelationship and interconnectedness; indeed the current wave 'new materialism' bears striking resemblance to, and could benefit from *indyamarra*.
>
> (Abdilla and Fitch 2017: 8)

Ngapulara Ngarngarnyi Wirra, then, offers an intimate experience of the inextricable connection of kinship and Country that is common across Indigenous knowledge systems more broadly.

At the gallery, even before entering the artwork, we are first met with a calling to the spirits, spoken in Adnyamathanha language by an Arraru Elder. This call creates a liminal threshold, designed to slow down the senses. We are then invited to walk into a darkened gallery space to view a six-metre high by two and a half metre wide box, projected on east and west sides with a gigantic, animated, glowing tree (Figure 14.2).

The image is constructed from clusters of tiny star-like dots coloured in greens, yellows, oranges, and reds against a stark black void. The branches and canopy of the tree are outlined in green, and a thick trunk leads to an elaborate root system in oranges and reds.

Figure 14.2 Ngapulara Ngarngarnyi Wirra (Our Family Tree), 2022.
Courtesy of Old Ways, New; Museum of Discovery (MOD), TopBunk Photography.

The tree image moves, slowly revolving and even spinning upside down. The amplified sounds of warm and cool winds lead this movement. As the tree spins, the image zooms in and out; sometimes, the entire tree shape is visible, and at other moments, we are inside the canopy or root system, with a close-up view of data points that are constantly rippling, swirling, and flowing. At times, it feels as though the tree is swallowing us.

We then move 'inside' the tree via entrances on opposing sides of the box. Here within lies a subversion. Across two counterposed LCD screens, vast, epic cinematic scenes of Country balance the tree's representation of Kin. Narrated in Adnyamathanha language by Goodes, *Arraru* and his Elder, *Mathari*, we hear a creation story that tells of the beginning of time. The dual video channels present the sacred places referenced in the creation story, within Adnyamathana Country.

In making *Ngapulara Ngarngarnyi Wirra,* the team drew on approximately 20 million data points, derived from a biometric performance management device used by the Australian Football League. Over the last years of Goodes' playing career, as he moved across sporting fields, training grounds and in personal life, his body was tracked and recorded. This biometric tracking also generated intimate, personal data. During a 2013 AFL match, Goodes reacted to a racist slur personally directed at him from a spectator in the stadium by pointing into the crowd. After which, from 2013 to 2015, Goodes was then targeted through repeated mass-booings on the field, regularly racially abused at matches and intensively vilified by the media. The booing continued to escalate, leading to a significant deterioration of Goodes' mental health and eventually forcing his retirement from the AFL.[3]

In *Ngapulara Ngarngarnyi Wirra*, Goodes' biometric data is repatriated and embedded into a volumetric data scan of the ancient river red gumtree, the Wirra. Surrounding the Wirra's data points, Goodes' data oscillates in swirls, embodying the ghost of the racism Goodes confronted throughout his career and drawing attention to histories of surveillance targeting Black bodies both in Australia and elsewhere. In light of this history, Abdilla felt a cultural responsibility to care for these datasets by symbolically repatriating the tracking data back to Country and kin. In doing so, the principles of relationality and rules of the kinship system guided the data visualisation, spatial design of the installation, and the direction for media within the artwork's video channels.

However, what were disparate elements during the creative development, but important to include due to cultural responsibilities, was a necessitated linking of the conceptual and creative elements to create a cohesive experience for the viewer. Machine learning interlinks the critically connected elements of Country and kin by training the data derived from the creation story, narrated by the Ararru and Mathari men, for a neural network model. The model was programmed according to the rules of warm and cool temperatures to produce a sonic output representing the two winds that form critical parts of the kinship system. The model's output formed the basis of the soundscape, connecting the movement of the Wirra with internal and external spaces, elements and cultural states.

From inside the Wirra, the distant sounds of the circulating wind soundscape seep through. Conversely, from outside the box upon which the Wirra is projected, the whispers of narration enhance the spatiality of this work. Utilising kinship rules to underpin the programming protocols and training of data and logic within the machine learning model enabled a divergent use of automation, enculturated within an Indigenous paradigm (rather than Western data protocols). As such, along with the centrality of the tree, the immersive soundscape of wind is an integral part of *Ngapulara Ngarngarnyi Wirra*. For Adnyamathanha people, the movement of Ararru and Mathari through the leaves of the trees carries the rhythm of their language. These interconnections flow between tree, root, air and atmosphere, animating dynamic exchanges of knowledge and kinship, protocols that are connected through the winds.

The oscillating movement of the digital tree is a manifestation of the life-giving properties of trees for all planetary life. In particular, the point cloud animation recalls the process of photosynthesis in which trees convert energy from the sun, along with water and nutrients drawn from their roots, into sugars, minerals and chemicals. Photosynthesis is described in detail by Natasha Myers as:

> [A] complex suite of electrochemical processes that spark energy gradients across densely folded membranes inside the symbiotic chloroplasts

> of green beings. Textbook diagrams familiar from high-school biology class are simplistic renderings of that utterly magical, totally cosmic alchemical process that tethers earthly plant life in reverent, rhythmic attention to the earth's solar source. Plants ... draw on the energy of the sun to rearrange the elements: they compose with hydrogen, water, oxygen, carbon dioxide, nitrogen, phosphorous, potassium, calcium, magnesium, zinc, sulfur, chlorine, boron, iron, copper.
>
> (2017: 124)

Where extractive colonial practices seek to separate minerals and elements into orders and categories, *Ngapulara Ngarngarnyi Wirra* suggests a dynamic and patterned interrelationship between technology, humans and nature. The tree acts as a convertor, in this case, changing the negative deployments of biometric data commonly used to police Indigenous bodies into an experience of repatriation and reconnection with Country and Kin.

In this way, the image of the tree in constant motion is not only a visualisation of important cultural connections, but a symbol for deep-time ancestral relations and knowledges held within Country. Moreover, this work models new possibilities for connecting and caring for Country when Indigenous-led protocols inform the development of AI (Abdilla et al. 2021; Lewis 2020). As the data points forming the Wirra move like water, it becomes evident that the flowing animation of the tree is not only a representation of photosynthetic activity, kinship connections and ancestral ties but also a dynamic visualisation of the cultural nuances that live in the very data from which it is constructed (Abdilla et al. 2021). Lines of connection run throughout and around the tree – movements of water, nutrients, energy, creatures, critters and organisms big and small in a symbiotic relation of reciprocity with the Wirra. These connections link the deep past of the 500-year-old Wirra, a being that has witnessed geographical changes, plant evolution and colonisation of culture and soil, with current movements of particles, energies and light that reflect the everywhen of past-present-future experiences of time. Fundamentally, the connections between language, time, space and matter that encompass *Ngapulara Ngarngarnyi Wirra* are cultural, elemental and radically everyday.

In contrast, images of more-than-human connection within Western theatre and performance remain remarkably static. Consider the twentieth century's archetypal Act 1 description of the tree in Samuel Beckett's *Waiting for Godot* (1965): 'A country road. A tree. Evening' (6). The lone tree, symbolically representative of war, conflict, nuclear fallout, slow violence or other images of climatic and genocidal devastation, has only recently been read in relation to ecological practice (Lavery and Finburgh 2015; Lavery 2018; Tsushima et al. 2023). Notably, Carl Lavery reads time and co-presence in Beckett through the metaphor of the garden, while Trish McTighe positions Beckett's tree in the context of colonial deforestation in Ireland (Lavery 2018; McTighe 2023). David Williams, in his epilogue for Lavery and Finburgh's *Rethinking the Theatre of the Absurd*, includes a brief description of a Beckett-parody by the muppets, *Waiting for Elmo*. Williams notes the 'astonishing' moment in which the tree reveals that it is sentient and begins to speak – complaining that nothing is happening, the play doesn't make sense and asking why couldn't they all perform *Oklahoma* instead (2015: 244). Yet, even in Williams' playful description of muppet-tree-agency, scant attention is given to the ways in which our engagement with Beckett's tree might be reimagined to foreground more-than-human connections on-stage.

At the beginning of Act 2, the tree in *Waiting for Godot* is shown to have 'four or five leaves', which Vladimir regards for an extended time (1965: 57). This evidence of tree life and, by association, the unseen to the human eye actions of photosynthesis reveals a tree capable of growth and regeneration. Is the beginning of Act 2 the beginning of a new season, or is the leaf growth an indication of greater soil health, water retention or increasing sunlight? Indeed, just as we read the mound of soil in *Happy Days* as a dynamic system made up of millions of biological agents and organisms ready to intermingle with Winnie's human form in our chapter on Ants, the tree in *Waiting for Godot* can also be read in relation to new understandings of tree communication and intelligence and the ways in which trees are intimately interconnected with other systems and beings (Kohn 2013).

Similar to Lavery's argument for Beckett's theatre as a *sensorium* rather than simply a 'space for looking', we are interested in the ways in which more-than-human elements start to enter and affect the enclosed biospherical spaces of theatre making and training (2018: 17). When we speak of rejecting the biosphere, we don't mean that all performances must take place outside or in the soil, but instead that future theatre makers and spectators pay more attention to creating intersections between beings and things rather than proliferating performances of human exceptionalism.[4] In this way, the tree in *Waiting for Godot* may continue to stand as a symbol, but it may also stand as a practical example of ecological connection.

While we have placed emphasis on four or five leaves as an indication of living systems within a performing climate, the roots and trunks of trees that appear long-dead might also be reconsidered as important links to past and present habitats. As part of the Monash University Museum of Art exhibition *Tree Story* in 2021, Indigenous Australian artist and scholar Brian Martin, Bundjalung, Kamilaroi and Muruwari exhibited his series: *Methexical Countryscapes*. Drawing on the Greek word *methexical*, Martin refers to a 'collaborative performative act that brings something into being' (2021). His works, large charcoal drawings of trees and plants inspired by site-visits to Country, present a participatory invitation to learn about understandings of Country from an Indigenous perspective.

In his drawings, Martin refigures and critiques the European artists' tendency for crisp, distanced and 'objective' depictions of harmonious landscape practices. Instead, Martin focuses on close-up viewpoints of tree forms, undergrowth and messy biological intermingling. In *Methexical Countryscape: Kamilaroi #10* (2017), the trunks of fallen trees on Kamilaroi Country lay horizontally across the image, dominating more than three-quarters of the composition, with a small patch of sky in the top-centre flanked by canopies of distant trees. The horizontal forms are pitted, hollowed-out trunks, stumps and branches. Organic plant material, leaves, bark, earth, lichen, and grasses seem to crawl through the fallen trees, forming a series of deeply textured layers. For Martin, the Indigenous relation to more-than-human agency is key. He confirms that 'an Indigenous view of the world looks at Country as its own [living] subjectivity' (Martin 2021: np), meaning that the zoomed-out perspective of European painters that depicted Australian landscapes with classical composition and distinct horizon lines is a romantic fantasy that removes the agency and messiness of the more-than-human in favour of a human-centric view.

In order to further unsettle these colonial fantasies of the landscape, Martin insists that his artworks be installed on the gallery floor and invites viewers to literally 'walk' on Country. In *Methexical Countryscape: Boon Wurrung #1* (2021), a gigantic tree, this time standing vertically, takes up the centre of the frame. The tree is from Boon Wurrung Country and is a sacred being also known as the Separation Tree – representing the boundaries of Country between the North-Eastern States of New South Wales and Victoria. In 2010 and 2013 the tree, which predated European colonisation, was vandalised and 'ring-barked' – meaning a significant part of Indigenous cultural heritage was desecrated and the tree's survival was jeopardised (Monument Australia 2023).

In Martin's rendering, the trunk of the Separation tree is in the foreground, with a background of layers of plant and tree life that completely overwhelm the composition. There is no visible horizon line and only a few specks of sky are visible. As Martin notes: 'In the Indigenous configuration of the world, when you're walking through the bush, you look down, you don't look toward the horizon' (Martin 2021). By situating this artwork on the gallery floor and explicitly encouraging viewers to walk across it, Martin refigures public engagement with his work and with Country.

We are not only invited to look down at our feet when we cross *Methexical Countryscape: Boon Wurrung #1*, we are drawn into the close-up intermingling of trees, leaves, grasses, earth and rocks. We are summoned to touch Boon Wurrung Country and consider the dynamic everyday existence of layers of biological beings, rather than looking towards a projected future. As Heather Davis reminds us, the etymology of the Greek term *Anthropos* is one who looks at the sky, which stands in contrast to the hummus that is connected to the earth and soil (2022: 51). Similarly, for Martin, settlers always have their sights set on the horizon, never grounded, looking to the future for more, with all the hubris that assumes (2021). In contrast, Abdilla references the importance of the horizon in *Ngapulara Ngarngarnyi Wirra* as a way of knowing where you are, anywhere in time and space. By reading the sky in relation *with* freshwater/saltwater/earth, we can deepen this connection through relationality.

The *Tree Story* exhibition featuring Martin's *Methexical Countryscapes* extended its invitation to collaborate and share ideas through a two-week education program, *Tree School* (2021). *Tree School* was led by Indigenous Australian practices of communal learning and knowledge creation, such as yarning circles and conversations, and structured around open-ended questions about 'caring for Country, learning from trees and storytelling', and engaging with alternative architectural, scholarly, artistic and educational practices (Monash University Museum of Art 2021a). The initial curators of the school, artists and architects Sandi Hilal and Alessandro Petti, and Yorta and Wurundjeri-Woiwurrung artist and educator Moorina Bonini, worked with Martin and Elder N'arweet Carolyn Briggs to bring the fallen trunk of an ancient 'scar-tree' from Boon Wurrung Country onto the Monash University campus as the physical and imaginary centre of the *Tree School* conversations.

The Boon Wurrung scar-tree reveals markings made by Indigenous people, including a part of the tree bark which was removed to make a shield. Importantly, the scar is not a sign of harmful practice but rather, in the words of Professor Jacinta Elston, 'an acknowledgement of the fact that the tree is healed. They didn't cut down the whole tree, they cut out a piece, they made the shield. The tree has healed itself' (Monash University Museum of Art 2021b). The journey of the scar-tree as it was relocated onto the university campus followed cultural protocols, including consultations with Boon Wurrung Elders such as Briggs, and the acknowledgement that the Wominjeka Djeembana Research Lab at Monash University was now the custodian of the tree for the 'purpose of teaching and learning'. As Martin welcomed visitors to *Tree School* he acknowledged the privilege of this custodianship and also paid respect to the tree as a living ancestor (Monash University Museum of Art 2021b).

In the beginning of this book, we made central our own acknowledgements and relations, including the ways in which we have tried to think-through and act in relation to what is both ours and not ours. We named our responsibility – as artists, parents, kin, teachers and learners – to take up custodianship for the place in which we live and foreground our care for the creatures, beings, elements, and stories that surround us. The stories and knowledges of trees, not as arborescent hierarchies but interconnected multispecies systems of relations, alchemy and ancient wisdom, offer a model of this care that is rarely reflected in Western theatrical practice. However, in writing about performing climates in an age of climate crisis, we envisage Western performance as an artform capable of grappling with its human-centric ways of conceiving, making and teaching. We reimagine a compostable artform embedded in the earth, in and of decay, in and of atmospheres, and in and outside of human-scale and human-perception.

We begin and end with the misplaced *Agonis*, whose roots are interacting with organisms and elements, whose leaves are rearranging energies and air beyond, then falling to the ground to rot and decay. We begin and end by acknowledging Country and the custodianship and teachings of the Dhudhuroa and Wayrurru peoples. We begin and end in relation to species, organisms, atmospheres and things: ants, ears, spiders, unicorns, rocks, plastics, mermaids, ice, death, roots, worms, trees. We begin and end with thanks to scholars of ecology and performance whose work have inspired us and First Nations colleagues who have invited us to share their knowledge and are often leading the way in ecological thinking. We seek to look more attentively at the ground beneath our feet. We aspire to move closer and farther still, to the microscopic and the macro, to cycles of connection, growth, decay, regeneration. We begin and end with performing climates beyond the biosphere. We end this work in the mess we began. We begin and end always.

Notes

1 We experienced *Ngapulara Ngarngarnyi Wirra* on Wurundjeri Country, in the Science Gallery Melbourne, Naarm, 2022.
2 Old Ways, New is an Indigenous-owned and led social enterprise based on Gadigal Country in Sydney; see oldwaysnew.com. The creative consultancy for *Ngapulara Ngarngarnyi Wirra (Our Family Tree)* was commissioned by the Museum of Discovery (MOD) to make the artwork for the exhibition *Invisibility*, launched in 2022.
3 The sustained racism directed at Goodes across 2013–2015 is the subject of two 2018 documentary films – *The Final Quarter* and *The Australian Dream*. In 2019, the Australian Football League issued an apology to Goodes for failing to support him during this period.
4 Lavery makes a similar point in his discussion of black-box theatres and site-specific performance (2018: 23).

References

Australia, Monument (2023). 'Separation Tree.' monumentaustralia.org.au/themes/government/colonial/display/32484-separation-tree-/, accessed 2 December 2023.

Abdilla, Angie and Robert Fitch (2017). 'Indigenous Knowledge Systems and Pattern Thinking: An Expanded Analysis of the First Indigenous Robotics Prototype Workshop'. *The Fibreculture Journal* 28(23): 1–14. www.fibreculturejournal.org, accessed 15 April 2024.

Abdilla, Angela, Megan Kelleher, Rick Shaw and Tyson Yunkaporta (2021). *Out of the Black Box: Indigenous Protocols for AI*. Melbourne: Deakin University.

Beckett, Samuel (1965). *Waiting for Godot: A Tragicomedy in Two Acts*. 2nd edn. London: Faber and Faber.

Davis, Heather (2022). *Plastic Matter*. Durham, NC: Duke University Press.

Harle, Josh, Angie Abdilla and Andrew Newman, eds. (2019). *Decolonising the Digital: Technology as Cultural Practice*. Sydney: Tactical Space Lab.

Kohn, Eduardo (2013). *How Forests Think: Towards an Anthropology Beyond the Human*. Berkley: University of California Press.

Lavery, Carl (2018). 'Ecology in Beckett's Theatre Garden: Or How to Cultivate the Oikos'. *Contemporary Theatre Review*. 28(1): 10–26.

Lavery, Carl and Clare Finburgh, eds. (2015). *Rethinking the Theatre of the Absurd: Ecology, the Environment and the Greening of the Modern Stage*. London: Bloomsbury.

Lewis, Jason Edward, ed (2020). *Indigenous Protocol and Artificial Intelligence Position Paper*. Honolulu, Hawaiʻi: The Initiative for Indigenous Futures and the Canadian Institute for Advanced Research (CIFAR).

Martin, Brian (2021). Interview. In *Tree Story: New Commissions*, video produced for Monash University Museum of Art. vimeo.com/521229001, accessed 13 November 2023.

Martin, Brian (2022). 'Methexical Countryscape Kamilaroi #10 2017.' Work description. Monash University Museum of Art, February. www.monash.edu/muma/education/deadly-learning/artists/brian-martin, accessed 14 December 2023.

McTighe, Trish (2023). 'Gestures of Helpless Compassion: Beckett's Eco-Poetics of Extinction'. In Michiko Tsushima, Yoshiki Tajiri and Mariko Hori Tanaka, eds, *Samuel Beckett and Catastrophe*. London: Palgrave MacMillan, pp. 193–213.

Methexical Countryscape: Boon Wurrung #1 (2021). Artwork by Brian Martin.

Methexical Countryscape: Kamilaroi #10 (2017). Artwork by Brian Martin.

Michiko, Tsushima, Yoshiki Tajiri and Mariko Hori Tanaka, eds. (2023). *Samuel Beckett and Catastrophe*. London: Palgrave MacMillan.

Monash University Museum of Art (2021a). 'The Tree School, Educational program (11-23 March 2021)'. https://www.monash.edu/muma/public-programs/previous/2022/tree-school, accessed 9 October 2024.

Monash University Museum of Art (2021b). *The Boon Wurrung Tree*. Video. vimeo.com/521716010, accessed 13 November 2023.

Munro, Kim (2022). 'A new exhibition explores invisible data, from facial algorithms to satellite tracking as a return to Country'. *The Conversation*, 22 February. theconversation.com/a-new-exhibition-explores-invisible-data-from-facial-algorithms-to-satellite-tracking-as-a-return-to-country-176567, accessed 21 August 2023.

Myers, Natasha (2017). 'Photosynthetic Mattering: Rooting into the Planthroposcene'. In Line Marie Thorsen, ed., *Moving Plants*. Næstved, Denmark: Rønnebæksholm, pp. 123–130.

Ngapulara Ngarngarnyi Wirra (Our Family Tree) (2022). Artwork by Old Ways, New.

Old Ways, New (2023). Work description. oldwaysnew.com/research-dev#spatial-awareness, accessed 14 December 2023.

The Australian Dream (2019). *Documentary Film, Dir. Daniel Gordon*. Australia: Good Thing Productions/Lorton Entertainment/Passion Pictures.

The Final Quarter (2019). *Documentary Film, Dir. Ian Darling*. Australia: Shark Island Productions.

Williams, David (2015). 'Epilogue: 'The Ruins of Time (I've Forgotten This Before)'. In Carl Lavery and Clare Finburgh, eds, *Rethinking the Theatre of the Absurd: Ecology, the Environment and the Greening of the Modern Stage*. London: Bloomsbury, pp. 241–250.

References

Abdilla, Angela, Megan Kelleher, Rick Shaw and Tyson Yunkaporta (2021). Out of the Black Box: Indigenous Protocols for AI. Report. Deakin University.

Abdilla, Angie and Robert Fitch (2017). 'Indigenous Knowledge Systems and Pattern Thinking: An Expanded Analysis of the First Indigenous Robotics Prototype Workshop'. *The Fibreculture Journal*. 28, 23 Jan: 1–14. www.fibreculturejournal.org, accessed 15 April 2024.

Abrahamsson, Christian and Sebastian Abrahamsson (2007). 'Cultural Geographies in Practice: In Conversation with the Body Conveniently Known as Stelarc'. *Cultural Geographies*. 14(2): 293–308.

Advisors, Dalberg, Wijnand de Wit and Nathan Bigaud (2019). *No Plastic in Nature: Assessing Plastic Ingestion from Nature to People*. Gland, Switzerland: World Wildlife Fund for Nature.

Alaimo, Stacy (2010). *Bodily Natures: Science, Environment, and the Material Self*. Bloomington: Indiana University Press.

Alaimo, Stacy (2014). 'Thinking as the Stuff of the World'. *O-Zone: A Journal of Object-Oriented Studies*. 1(1): 13–21.

Alaimo, Stacy (2016). *Exposed: Environmental Politics and Pleasures in Posthuman Times*. Minneapolis: University of Minnesota Press.

Andersen, Hans Christian (2017). *Hans Christian Andersen's Fairytales*, ed. JH Stickney. Minneapolis: Lerner Publishing Group.

Australian Museum (2021). 'Thylacine'. www.australian.museum/learn/australia-over-time/extinct-animals/the-thylacine/, accessed 9 November 2023.

Australian National Herbarium (2013). 'Australian Fungi'. www.anbg.gov.au/fungi/mycorrhiza.html, accessed 9 November 2023.

Barad, Karen (2007). *Meeting the Universe Halfway: Quantum Physics and the Entanglement of Matter and Meaning*. Durham, NC: Duke University Press.

Barry, Colleen (2023). 'Venice and Its Lagoon Again Escape Inclusion on UNESCO List of Heritage Sites in Danger'. *AP News*, 15 September. www.apnews.com/article/venice-unesco-world-heritage-downgrade-threat-45a7e0d51069c9f06e79b69bd8bd1ed3, accessed 10 November 2023.

Bartholomew, Jeff (2023). 'Lego Abandons Effort to Make Bricks from Recycled Plastic Bottles'. *The Guardian*, 25 September. www.theguardian.com/lifeandstyle/2023/sep/24/lego-abandons-effort-to-make-bricks-from-recycled-plastic-bottles, accessed 8 December 2023.

Beckett, Samuel (1938). *Murphy*. London: Calder & Boyars.

Beckett, Samuel (1965). *Waiting for Godot: A Tragicomedy in Two Acts*. 2nd edn. London: Faber and Faber.

Beckett, Samuel (1990). *The Complete Dramatic Works*. London: Faber & Faber.

Bellew, Patrick and Meredith Davey (2012). *Green House, Green Engineering: Environmental Design at Gardens by the Bay*. Novato, CA: Oro Editions.
Bennett, Jane (2010). *Vibrant Matter: A Political Ecology of Things*. Durham, NC: Duke University Press.
Biosphere (2022). Film, dir. Mel Eslyn. Duplass Brothers Productions, United States.
Birch, Tony (2016). 'Climate Change, Recognition and Social Place-Making'. In Eve Vincent and Timothy Neale, eds, *Unstable Relations: Indigenous People and Environmentalism in Contemporary Australia*. Crawley: UWAP Scholarly, pp. 356–383.
Birch, Tony (2019a). 'Walking and Being'. *Meanjin Journal*, Summer 2019. https://meanjin.com.au/essays/walking-and-being/
Birch, Tony (2019b). *HumanNature: Connection and Cooperation in a Time of Climate Change*. Australian Museum HumanNature Podcast Series, 14 May 2019.
Björk (2011). *Biophilia*. Recorded Work, App, Exhibition and Documentary Project.
Björk (2017). *Utopia*. Recorded Work. One Little Independent Records, United Kingdom.
Björk (2019). *Cornucopia*. Theatrical Concert.
Björk (2022). *Fossora*. Recorded Work. One Little Independent Records, United Kingdom.
Björkman, E (1970). 'Forest Tree Mycorrhiza: The Conditions for Its Formation and the Significance for Tree Growth and Afforestation'. *Plant Soil*. 32: 589–610.
Blake, Barry J. and Julie Reid (2002). 'The Dhudhuroa Language of Northeastern Victoria: A Description Based on Historical Sources'. *Aboriginal History*. 26: 177–210.
Blakemore, Rob (2019). 'Australian Earthworms: Factsheet'. Australian Museum, 25 February. www.australian.museum/learn/animals/worms/australian-earthworms, accessed 11 December 2023.
Bleach, Lucy (2015). *underground*. Work Description. www.lucybleach.com.au/lucy-bleach/selected-works/underground/, accessed 9 November 2023.
Bleeker, Maaike and Iris van der Tuin (2014). 'Science in the Performance Stratum: Hunting for Higgs and Nature as Performance'. *International Journal of Performance Arts and Digital Media*. 10(2): 232–245.
Body of Art (2013). Short Film, dir. Iona Reto. vimeo.com/60363446, accessed 1 June 2022.
Bramley, Ellie Violet (2018). 'Losing Its Sparkle: The Dark Side of Glitter'. *The Guardian*, 22 January. www.theguardian.com/fashion/2018/jan/21/losing-its-sparkle-the-dark-side-of-glitter, accessed 28 November 2023.
Bringhurst, Robert and Jan Zwicky (2018). *Learning to Die: Wisdom in the Age of Climate Crisis*. Regina, Saskatchewan: University of Regina Press.
Buranyi, Stephen (2023). '"We are Just Getting Started": The Plastic-Eating Bacteria That Could Change the World'. *The Guardian*, 28 September. www.theguardian.com/environment/2023/sep/28/plastic-eating-bacteria-enzyme-recycling-waste, accessed 8 December 2023.
Cai, Ming, Yang Lin, Yin-long Luo et al. (2015). 'Extraction, Antimicrobial, and Antioxidant Activities of Crude Polysaccharides from the Wood Ear Medicinal Mushroom *Auricularia auricula-judae* (Higher Basidiomycetes)'. *International Journal of Medicinal Mushrooms* 17(6): 591–600.
Cardoso, Maria Fernanda (2018a). 'On the Origins of Art I–II, 2016'. Work Description, *Museum of Contemporary Art*. www.mca.com.au/artists-works/works/2018.13/, accessed 3 November 2023.
Cardoso, Maria Fernanda (2018b). 'On the Origins of Art – Actual Size II Maratus Splendens, 2016'. Work Description, *Museum of Contemporary Art*. www.mca.com.au/artists-works/works/2018.15/, accessed 3 November 2023.
Carreon, Erika (2024). 'Tropical Futurity'. Ph.D. Thesis, University of Melbourne.

Carrington, Damian (2020). 'Stop Blowing Up Bombs on Sea Floor, Say Whale Campaigners'. *The Guardian*, 3 November. www.theguardian.com/environment/2020/nov/03/stop-blowing-up-bombs-on-sea-floor-say-whale-campaigners, accessed 10 November 2023.
Carton, Wim (2020). 'Carbon Unicorns and Fossil Futures: Whose Emission Reduction Pathways Is the IPCC Performing?'. In JP Sapinski, Holly Jean Buck and Andreas Malm, eds, *Has It Come to This? The Pitfalls and Promises of Geoengineering on the Brink*. New Brunswick, NJ: Rutgers University Press, pp. 34–49.
Catts, Oron and Ionat Zurr (2008). 'The Art of the Semi-Living and Partial Life: Extra Ear – 1⁄4 Scale'. In Melentie Pandilovski, ed, *Art in the Biotech Era*. Adelaide: Experimental Art Foundation, pp. 140–147.
Cervera, Felipe (2017a). 'Planetary Performance Studies'. *Global Performance Studies*. 1(1). https://doi.org/10.33303/gpsv1n1a3
Cervera, Felipe (2017b). 'The Planet as Method'. www.felipecervera.me/the-planet-as-method, accessed 9 November 2023.
Chang, Chia-ju (2016). 'Wasted Humans and Garbage Animals: Deadly Transcorporeality and Documentary Activism'. In Rayson K Alex and S Susan Deborah, eds, *Ecodocumentaries: Critical Essays*. London: Palgrave, pp. 95–114.
Cheung, Helier (2020). 'What Does Trump Actually Believe on Climate Change?' *BBC News*, 23 January. www.bbc.com/news/world-us-canada-51213003, accessed 19 October 2023.
Chevalier, Clarissa (2021). 'Eco-Phenomenology and the Maintenance of Eco Art: Agnes Denes's *A Forest for Australia*'. *Australian and New Zealand Journal of Art*. 21(2): 273–290.
Chiang, C, S Mahalingam and JA Flaws (2017). 'Environmental Contaminants Affecting Fertility and Somatic Health'. *Seminars in Reproductive Medicine*. 35(3): 241–249.
Churchill, Caryl (2000). *Far Away*. London: Nick Hern Books.
Churchill, Caryl (2016). *Escaped Alone*. London: Nick Hern Books.
Cixous, Hélène (1984). 'Aller à la mer', trans. Barbara Kerslake. *Modern Drama*. 27(4): 546–548.
Clark, Jane and Luke Hortle (2022). 'Spider and Stars'. Content Available on the Museum of Old and New Art Visitor App.
Collins, Rebecca (2021). 'Aural Attunement Is a Form of Performative Writing Used to Write About Nonhuman Affective Relations in Emma Bennett's *Slideshow Birdshow* (2013)'. *Ambiances*. 2: 1–20.
Cormick, Hanna (2020). *The Mermaid*. Performance Work, Sydney Festival, 17–19 January.
Cormick, Hanna (2021). Interviewed by Lara Stevens and Denise Varney, 26 February, Zoom.
Country Needs People website (n.d.). https://www.countryneedspeople.org.au/about, accessed 23 April 2024.
Cramer, Maria (2021). 'Leaky Gas Pipeline Sparks an Inferno in the Gulf of Mexico'. *The New York Times*, 3 July. www.nytimes.com/2021/07/03/world/americas/eye-fire-gulf-mexico.html, accessed 10 November 2023.
Crew, Bec (2016). 'The Robots Sent Into Fukushima Have "Died"'. *Science Alert*, 11 March. www.sciencealert.com/the-robots-sent-into-fukushima-have-died, accessed 7 November 2023.
Crocodile Dundee (1986). Film, dir. Peter Faiman. Rimfire Films, Australia.
Curry, JP and O Schmidt (2007). 'The Feeding Ecology of Earthworms: A Review'. *Pedobiologia*. 50(2007): 463–477.
Cut the Sky (2016). Performance Work, Marrugeku.
D'Cruz, Glenn (2014). 'Things I Know About Geminoid F, or What I Think About When I Think About Android Theatre'. *Australasian Drama Studies*. (65): 272–288.

Darwin, Charles (1881). *The Formation of Vegetable Mould, through the Action of Worms, with Observations on Their Habits*. London: John Murray.
Dastogeer, KMG et al. (2020). 'Plant Salinity Tolerance Conferred by Arbuscular Mycorrhizal Fungi and Associated Mechanisms: A Meta-Analysis'. *Frontiers in Plant Science*. 11: 588550.
Davis, Heather (2022). *Plastic Matter*. Durham, NC: Duke University Press.
Davis, Tracy C (2019). 'How Do You Know a Mermaid When You See One? How Do You See a Mermaid When You Know One?. *Theatre Journal*. 71(3): 257–288.
de Beauvoir, Simone (2009). *The Second Sex*, trans. Constance Borde and Sheila Malovany-Chevallier. New York: Vintage Books.
de Vietri, Gabrielle (2019). *Maps of Gratitude, Cones of Silence and Lumps of Coal*. Interactive Work. www.gabrielledevietri.net/maps-of-gratitude
Déchery, Chloé and Martin Welton, eds (2020). 'Staging Atmospheres: Theatre and the Atmospheric Turn'. *Ambiances* 1: Special Issue.
Déchery, Chloé and Martin Welton, eds (2021). 'Staging Atmospheres: Theatre and the Atmospheric Turn'. *Ambiances* 2: Special Issue.
Deleuze, Gilles and Félix Guattari (2013). *A Thousand Plateaus: Capitalism and Schizophrenia*, trans. Brian Massumi. London: Bloomsbury.
Denes, Agnes (1996). *Tree Mountain*. Artwork.
Denes, Agnes (1998). *A Forest for Australia*. Artwork.
Denes, Agnes (2023). Personal Website. www.agnesdenesstudio.com, accessed 9 November 2023.
Detail (2019). Installation Work by Kris Verdonck. www.atwodogscompany.org/en/projects/detail/.
Devlin, Hannah (2018). 'Scientists Discover Spider Species That Feeds Its Young Milk'. *The Guardian*, 30 November. www.theguardian.com/world/2018/nov/29/scientists-discover-spider-species-that-feeds-its-young-milk, accessed 3 November 2023.
Diamanti, Jeff (2023). *Climate and Capital in the Age of Petroleum: Locating Terminal Landscapes*. London: Bloomsbury.
Donald, Dwayne (2012). 'Indigenous Métissage: A Decolonizing Research Sensibility'. *International Journal of Qualitative Studies in Education*. 25(5): 533–555.
Duets on Ice (1974). Performance Work by Laurie Anderson.
Eckersall, Peter (2015). 'Towards a Dramaturgy of Robots and Object-Figures'. *TDR: The Drama Review*. 59(3): 123–131.
Eckersall, Peter (2020). 'The Poor Old Sun: A Note on Kris Verdonck's DETAIL'. In Peter Eckersall and Kristof van Baarle, eds, *Machine Made Silence: The Art of Kris Verdonck*. Aberystwyth: Performance Research Books, pp. 89–91.
Eckersall, Peter (2024). *Dramaturgy to Make Visible: The Legacies of New Dramaturgy for Politics and Performance in Our Times*. London: Routledge.
Eckersall, Peter and Eddie Paterson (2011). 'Slow Dramaturgy: Renegotiating Politics and Staging the Everyday'. *Australasian Drama Studies*. (58): 178–192.
Eckersall, Peter, Helena Grehan and Edward Scheer (2017). *New Media Dramaturgy: Performance, Media and New Materialism*. London and New York: Palgrave MacMillan.
Efi, Tui Ātua Tupua Tamasese Tupuola Tufuga (2019). 'Prelude: Climate Change and the Perspective of the Fish'. In Tony Crook and Peter Rudiak, eds, *Pacific Climate Cultures: Living Climate Change in Oceania*, Warsaw: De Gruyter, pp. ix–xiii.
Engelmann, Sasha (2016). 'Social Spiders and Hybrid Webs at Studio Tomás Saraceno'. *Cultural Geographies*. 24(1): 161–169.
Etchells, Tim (2008). *Spectacular* 'Programme Note'. Forced Entertainment. www.forcedentertainment.com/spectacular-programme-note-by-tim-etchells, accessed 7 December 2023.

Etchells, Tim (2011). *Tomorrow's Parties* 'Programme Note'. Forced Entertainment. www.forcedentertainment.com/tomorrows-parties-programme-note-by-tim-etchells/, accessed 15 December 2023.

Explore Beechworth (2017). 'Yeddonba Aboriginal Cultural Site'. www.explorebeechworth.com.au/yeddonba, accessed 5 December 2023.

Extinction Rebellion (2021). 'Nowhere Is Safe'. *Global Newsletter*. (53), 16 July. rebellion.global/blog/2021/07/16/global-newsletter-53/, accessed 14 December 2023.

Fantastic Fungi (2019). Documentary Film, dir. Louie Schwartzberg. Moving Art, United States.

Fensham, Rachel, Eddie Paterson and Paul Rae, eds (2018). 'On Climates'. *Themed Issue, Performance Research*. 23(3).

Finley, Karen (2018). *Grabbing Pussy*. New York: OR Books.

Finnigan, David (2023). 'Falling in Love with Microplastic'. Personal Website, 7 May. davidfinnigan.substack.com/p/falling-in-love-with-microplastic, accessed 7 November 2023.

First Peoples State Relations (2021). 'Fact Sheet: Aboriginal Stone Arrangements'. www.firstpeoplesrelations.vic.gov.au/fact-sheet-aboriginal-stone-arrangements, accessed 9 November 2023.

Fletcher, Michael-Shawn et al. (2022). 'Healthy Country'. *Indigenous Knowledge Institute* https://indigenousknowledge.unimelb.edu.au/Research/themes/2021-2022-themes/healthy-country, accessed 23 April 2024.

Forced Entertainment (2008). *Spectacular.* Performance Work.

Forced Entertainment (2011). *Tomorrow's Parties*. Performance Work.

French, Jeffrey and Sarah Tessendorf (2018). 'Does Cloud Seeding Work? Scientists Watch Ice Crystals Grow Inside Clouds to Find Out'. The Conversation, 14 March. www.theconversation.com/does-cloud-seeding-work-scientists-watch-ice-crystals-grow-inside-clouds-to-find-out-90903, accessed 12 October 2023.

Frew, Adam et al. (2023). 'Hidden Carbon: Fungi and Their 'Necromass' Absorb One-Third of the Carbon Emitted by Burning Fossil Fuels Every Year'. *The Conversation*, 6 June. www.theconversation.com/hidden-carbon-fungi-and-their-necromass-absorb-one-third-of-the-carbon-emitted-by-burning-fossil-fuels-every-year-206674, accessed 9 November 2023.

Furlong, Claire, Nalika S Rahapaksha, Kevin R Butt and Walter T Gibson (2017). 'Is Composting Worm Availability the Main Barrier to Large-Scale Adoption of Worm-Based Organic Waste Processing Technologies?' *Journal of Cleaner Production*. 164: 1026–1033.

Furuhata, Yuriko (2022). *Climatic Media: Transpacific Experiments in Atmospheric Control*. Durham and London: Duke University Press.

Gay'Wu Group of Women (2019). *Songspirals: Sharing Women's Wisdom of Country through Songlines*. Sydney: Allen & Unwin.

George, Cherian (2000). *Singapore, the Air-Conditioned Nation: Essays on the Politics of Comfort and Control, 1990–2000*. Singapore: Landmark Books.

Girard, Madeline B, Michael M Kasumovic and Damian O Elias (2011). 'Multi-Modal Courtship in the Peacock Spider, *Maratus Volans*'. (O.P.-Cambridge, 1874). *PLoS One*. 6(9): e25390.

Goodall, Jane (2013). 'Natural Aesthetics: Formations of Gender in the Work of Maria Fernanda Cardoso'. *Artlink*. 33(3): 52–55.

Graham, Mary (1999). 'Some Thoughts about the Philosophical Underpinnings of Aboriginal Worldviews'. *Worldviews: Environment, Culture, Religion*. 3: 105–118.

Graham, Mary (2009). 'Understanding Human Agency in Terms of Place: A Proposed Aboriginal Research Methodology'. *PAN: Philosophy Activism Nature*. 6: 71–78.

Greening the West (2023). 'A Forest for Australia – Altona Treatment Plant'. www.greeningthewest.org.au/projects/forest-australia-altona-treatment-plant/, accessed 9 November 2023.

Grehan, Helena (2015). 'Actors, Spectators, and "Vibrant" Objects: Kris Verdonck's *ACTOR #1*'. *TDR: The Drama Review*. 59(3): 132–139.

Grehan, Helena (2018). 'First Nations Politics in a Climate of Refusal'. *Performance Research: On Climates*. 23(3): 7–12.

Gulantang! A Children's Musical (2011). Performance Work, Cope/with/Land Theatre Company.

Hadingham, Evan (1975). *Circles and Standing Stones: An Illustrated Exploration of Megalith Mysteries of Early Britain*. New York: Walker.

Hamilton, Jennifer Mae and Astrida Neimanis (2018). 'Composting Feminisms and Environmental Humanities'. *Environmental Humanities*. 10(2) November: 501–527.

Haraway, Donna (1988). 'Situated Knowledges: The Science Question in Feminism and the Privilege of Partial Perspective'. *Feminist Studies*. 14(3): 575–599.

Haraway, Donna (1991). *Simians, Cyborgs, and Women: The Reinvention of Nature*. New York: Routledge.

Haraway, Donna (2017). 'Symbiogenesis, Sympoiesis, and Art Science Activisms for Staying with the Trouble'. In Anna Lowenhaupt Tsing, Heather Swanson, Elaine Gan and Nils Bubandt, eds, *Arts of Living on a Damaged Planet*. Minneapolis: University of Minnesota Press, pp. M25–M50.

Haraway, Donna J. (2003). *The Companion Species Manifesto: Dogs, People, and Significant Otherness*, vol. 1. Chicago: Prickly Paradigm Press.

Harle, Josh, Angie Abdilla and Andrew Newman, eds. (2019). *Decolonising the Digital: Technology as Cultural Practice*. Sydney: Tactical Space Lab.

Hau'ofa, Epeli (1994). 'Our Sea of Islands'. *The Contemporary Pacific*. 6(1): 147–162.

Hiebert, Ted (2023). 'Trust Fall: Rubbing Rocks and the Irreconcilability of Objects'. *Performance Research*. 27(2): 16–24.

High Life (2018). Film, dir. Claire Denis. Andrew Lauren Productions/Pandora Filmproduktion/Alcatraz Films/The Apocalypse Films Company/Madants, France.

Hirata, Oriza (2010). *Sayonara*. Performance Work, Seinendan Theatre Company.

Hodge, Gabrielle (2020). 'The Ideology of Communication Practices Embedded in an Australian Deaf/Hearing Dance Collaboration'. In Annelies Kusters, Mara Green, Erin Moriarty and Kristin Snoddon, eds, *Sign Language Ideologies in Practice*. Berlin: De Gruyter Mouton, pp. 59–82.

Holmes, Tracey (2022). 'Adam Goodes's AFL Data, Its Entanglement with Aboriginal Knowledge Systems, and the Sacred Wirra That's Protecting It'. *ABC News*, 28 May. www.abc.net.au/news/2022-05-28/adam-goodes-afl-data-set-sacred-wirra-tracker-data-project/101097190, accessed 8 July 2023.

Hoover, Elizabeth (2017). *The River Is in Us: Fighting Toxics in a Mohawk Community*. Minneapolis: University of Minnesota Press.

Houlton, Sarah (2013). 'Unicorns Don't Exist'. *Chemistry & Industry*. 77(8): 23.

Ibsen, Henrik (2012). *A Doll's House*, trans. Simon Stephens. London: Methuen Drama.

Ice Watch (2014). Performance Work by Olafur Eliasson and Minik Rosing.

i-Land X-isle (2012). Performance Work by Latai Taumoepeau.

Ingegnoli, Vittorio, Francesco Lombardo and Giuseppe La Torre, eds (2022). *Environmental Alteration Leads to Human Disease: A Planetary Health Approach*. Cham, Switzerland: Springer.

Ingwersen, Moritz (2017). '*Cybernethics*: Thinking Bodies and Boundaries through Science'. In Anne Waldschmidt, Hanjo Berressem and Moritz Ingwersen, eds, *Culture – Theory – Disability: Encounters between Disability Studies and Cultural Studies*. Bielefeld, Germany: Transcript, pp. 160–169.

Interstellar (2014). Film, dir. Christopher Nolan. Paramount Pictures/Warner Bros. Pictures/Legendary Pictures/Syncopy/Lynda Obst Productions, United States.

Iscoe, Adam (2022). 'Arachnophile'. *The New Yorker*. 98(6), 28 March: 22–23. *It's Still Life: Listening to Composts Still Life: Listening to Compost* (2019). Art installation, Out-of-Sync.

Jestrovic, Silvija (2004). 'From the Ice Cube Stage to Simulated Reality: Place and Displacement in Laurie Anderson's Performances'. *Contemporary Theatre Review*. 14(1): 25–37.

Jones, Andrew Merion (2012). 'Living Rocks: Animacy, Performance and the Rock Art of the Kilmartin Region, Scotland'. In Andrew Cochrane and Andrew Merion Jones, eds, *Visualising the Neolithic*. Oxford: Oxbow Books, pp. 79–88.

Jongen, Marc (2011). 'On Anthropospheres and Aphrogrammes: Peter Sloterdijk's Thought Images of the Monstrous'. *Humana.Mente Journal of Philosophical Studies*. 18: 199–219.

Kelly, Matthew W., Doris K. Wu, Arthur N. Popper and Richard R. Fay, eds (2005). *Development of the Inner Ear*. New York: Springer.

Klein, Naomi (2007). *The Shock Doctrine: The Rise of Disaster Capitalism*. London: Allen Lane.

Klinsukon, C, S Lumyong, TW Kuyper et al. (2021). 'Colonization by Arbuscular Mycorrhizal Fungi Improves Salinity Tolerance of Eucalyptus (*Eucalyptus camaldulensis*) Seedlings'. *Scientific Reports*. 11: 4362.

Kohn, Eduardo (2013). *How Forests Think: Towards an Anthropology Beyond the Human*. Berkley: University of California Press.

Kruszelnicki, Karl (2006). 'Mouse with Human Ear'. *ABC Science,* Australian Broadcasting Corporation, 2 June. www.abc.net.au/science/articles/2006/06/02/1644154.htm, accessed 5 July 2023.

Kwaymullina, Ambelin (2018). 'You Are on Indigenous Land: Ecofeminism, Indigenous Peoples and Land Justice'. In Lara Stevens, Peta Tait and Denise Varney, eds, *Feminist Ecologies: Changing Environments in the Anthropocene*. London: Palgrave Macmillan, pp. 193–208.

Laird, Tessa (2018). 'In the Beginning There Was the Worm: Animal Voice Beyond the Verbal'. *Artlink*. 38(1): 58–65.

Langton, Marcia (1995). 'The European Construction of Wilderness'. *Wilderness News*, Summer 1995–1996: 16–17.

Latour, Bruno (2018). *Down to Earth: Politics in the New Climatic Regime*, trans. Catherine Porter. Cambridge: Polity Press.

Lavers, Chris (2009). *The Natural History of Unicorns*. New York: William Morrow.

Lavery, Carl (2018). 'Ecology in Beckett's Theatre Garden: Or How to Cultivate the Oikos'. *Contemporary Theatre Review*. 28(1): 10–26.

Lavery, Carl, ed (2019). *Performance and Ecology: What Can Theatre Do?* London: Routledge.

Lavery, Carl and Clare Finburgh, eds (2015). *Rethinking the Theatre of the Absurd: Ecology, the Environment and the Greening of the Modern Stage*. London: Bloomsbury.

Law, Benjamin (2023). 'Björk Cornucopia Review – an Electrifying Pop Concert, Art Installation and Opening Ceremony Rolled Into One'. *The Guardian*, 4 March. www.theguardian.com/music/2023/mar/04/bjork-cornucopia-review-an-electrifying-pop-concert-art-installation-and-opening-ceremony-rolled-into-one, accessed 9 November 2023.

Leane, Elizabeth (2022). 'Ice Islands of the Anthropocene: The Cultural Meaning of Antarctic Bergs'. In Russell McDougall, John C. Ryan and Pauline Reynolds, eds, *Postcolonial Literatures of Climate Change*. Leiden: Brill, pp. 361–382.

Leane, Elizabeth, Carolyn Philpott and Matt Delbridge (2020). 'Performing Ice: Histories, Theories, Contexts'. In Carolyn Philpott, Elizabeth Leane and Matt Delbridge, eds, *Performing Ice*. Cham, Switzerland: Palgrave Macmillan, pp. 1–26.
Lee, Aileen (2013). 'Welcome to The Unicorn Club: Learning from Billion-Dollar Startups'. *TechCrunch*, 3 November. www.techcrunch.com/2013/11/02/welcome-to-the-unicorn-club/, accessed 19 October 2023.
Lego Breakfast (2020). Video by tomosteen, 17 July. www.youtube.com/watch?v=1RMKR8h03iw, accessed 16 October 2023.
Liang, F, X Huo, W Wang et al. (2020). 'Association of Bisphenol A or Bisphenol S Exposure with Oxidative Stress and Immune Disturbance Among Unexplained Recurrent Spontaneous Abortion Women'. *Chemosphere*. 257: 127035.
Liboiron, Max (2021). *Pollution Is Colonialism*. Durham, NC: Duke University Press.
Lorenzutti, Gregory (2022a). 'Director's Note'. Credits for *Theatrum Botanicum*. https://finearts-music.unimelb.edu.au/showcase/theatrum-botanicum/theatrum-botanicum-credits, accessed 2 October 2023.
Lorenzutti, Gregory (2022b). 'Microorganism and Macro-Movement: Soil & Food Cultivation for a Dance Ecology'. Presentation for Performance Studies International Conference #27. www.gregorylorenzutti.com/germinations/blog-post-title-one-86mrb-z4fwc-s6xfb-shh75-5yk6n-3x9ya-4s26l-x8sr9-p85zt-2k55t-kc2r5-gzw9w, accessed 4 September 2023.
Lowenfels, Jeff (2017). *Teaming with Fungi*. Portland, OR: Timber Press.
Luo, Yangchao, Gang Chen, Bo Li et al. (2009). 'Evaluation of Antioxidative and Hypolipidemic Properties of a Novel Functional Diet Formulation of *Auricularia auricula* and Hawthorn'. *Innovative Food Science and Emerging Technologies*. 10(2): 215–221.
Maiquez, Reagan, Dennis Gupa and Kulay Labitigan (2018). 'Between Theatre and the Environment: The Experience of Cope/With/Land Theatre Co'. *Global Performance Studies*. 1(2).
Mangioni, Talei Luscia (2021). 'Confronting Australian Apathy: Latai Taumoepeau and the Politics of Performance in Pacific Climate Stewardship'. *The Contemporary Pacific*. 33(1): 32–62.
Mapoung, Sariya, Sonthaya Umsumarng, Warathit Semmarath et al. (2021). 'Skin Wound-Healing Potential of Polysaccharides from Medicinal Mushroom *Auricularia auricula-judae* (Bull.)'. *Journal of Fungi*. 7: 247.
Martin, Brian (2021). Interview. In *Tree Story: New Commissions*, video produced for Monash University Museum of Art. vimeo.com/521229001, accessed 13 November 2023.
Martin, Brian (2022). 'Methexical Countryscape Kamilaroi #10 2017'. Work Description. Monash University Museum of Art, February. www.monash.edu/muma/education/deadly-learning/artists/brian-martin, accessed 14 December 2023.
McCarthy, Terry (1999). 'Lee Kuan Yew'. *Time Magazine*, 23 August. http://content.time.com/time/world/article/0,8599,2054444,00.html#ixzz2lzv7Abbn, accessed 29 November 2013.
McCurry, Justin (2017). 'Fukushima Nuclear Cleanup Falters Six Years After Tsunami'. *The Guardian*, 9 March. www.theguardian.com/world/2017/mar/09/fukushima-nuclear-cleanup-falters-six-years-after-tsunami, accessed 1 September 2023.
McDougall, Russell, John C Ryan and Pauline Reynolds, eds (2022). *Postcolonial Literatures of Climate Change*. Leiden: Brill.
McGrath, Ann and Laura Rademaker (2023). 'The Languages and Temporalities of "Everywhen" in Deep History'. In Ann McGrath, Laura Rademaker and Jakelin Troy, eds, *Everywhen: Australia and the Language of Deep History*. Sydney: UNSW Press, pp. 1–36.
McKenzie, Jon (2001). *Perform or Else: From Discipline to Performance*. New York: Routledge.

McTighe, Trish (2023). 'Gestures of Helpless Compassion: Beckett's Eco-Poetics of Extinction'. In Michiko Tsushima, Yoshiki Tajiri and Mariko Hori Tanaka, eds, *Samuel Beckett and Catastrophe*. London: Palgrave MacMillan, pp. 193–213.

Methexical Countryscape: Boon Wurrung #1 (2021). Artwork by Brian Martin.

Methexical Countryscape: Kamilaroi #10 (2017). Artwork by Brian Martin.

Michiko, Tsushima, Yoshiki Tajiri and Mariko Hori Tanaka, eds (2023). *Samuel Beckett and Catastrophe*. London: Palgrave MacMillan.

Milanović, M, N Milošević, N Milić et al. (2023). 'Food Contaminants and Potential Risk of Diabetes Development: A Narrative Review'. *World Journal of Diabetes* 14(6): 705–723.

Missingham, Racheal (2021). 'Australia's Deaf Theatre: Past, Present and Future'. Master of Philosophy Thesis, Queensland University of Technology.

Monash University Museum of Art (2021a). 'The Tree School, Educational Program (11–23 March 2021)'. https://www.monash.edu/muma/public-programs/previous/2022/tree-school, accessed 13 November 2023.

Monash University Museum of Art (2021b). *The Boon Wurrung Tree*. Video. vimeo.com/521716010, accessed 13 November 2023.

Montgomery, David R (2008). *Dirt: The Erosion of Civilizations*. Berkeley: University of California Press.

Monument Australia (2023). 'Separation Tree'. monumentaustralia.org.au/themes/government/colonial/display/32484-separation-tree-/, accessed 2 December 2023.

Moreton-Robinson, Aileen (2000). *Talkin' up to the White Woman: Aboriginal Women and Feminism*. *St*. Lucia: University of Queensland Press.

Moreton-Robinson, Aileen (2013). 'Towards an Australian Indigenous Women's Standpoint Theory: A Methodological Tool'. *Australian Feminist Studies*. 28(78): 331–347.

Moreton-Robinson, Aileen (2015). *The White Possessive: Property, Power, and Indigenous Sovereignty*. Minneapolis: University of Minnesota Press.

Munro, Kim (2022). 'A New Exhibition Explores Invisible Data, from Facial Algorithms to Satellite Tracking as a Return to Country'. *The Conversation*, 22 February. theconversation.com/a-new-exhibition-explores-invisible-data-from-facial-algorithms-to-satellite-tracking-as-a-return-to-country-176567, accessed 21 August 2023.

Murphy, Michelle (2017). 'Alterlife and Decolonial Chemical Relations'. *Cultural Anthropology*. 32(4): 494–503.

Murphy, Michelle (2018). 'Against Population, towards Alterlife'. In Adele E Clarke and Donna Haraway, eds, *Making Kin Not Population*. Chicago: Prickly Paradigm Press, pp. 101–124.

Myers, Natasha (2015). 'Edenic Apocalypse: Singapore's End-of-Time Botanical Tourism'. In Heather Davis and Etienne Turpin, eds, *Art in the Anthropocene: Encounters Among Aesthetics, Politics, Environments and Epistemologies*. London: Open Humanities Press, pp. 31–42.

Myers, Natasha (2017a). 'From the Anthropocene to the Planthroposcene: Designing Gardens for Plant/People Involution'. *History and Anthropology*. 28(3): 297–301.

Myers, Natasha (2017b). 'Photosynthetic Mattering: Rooting into the Planthroposcene'. In Line Marie Thorsen, ed, *Moving Plants*. Næstved, Denmark: Rønnebæksholm, pp. 123–130.

Nagengast, Bernard A (2009). 'Was Ice Making John Gorrie's Greatest Legacy?'. *ASHRAE Transactions*. 115(1): 122–129.

National Indigenous Australians Agency (2023). 'Referendum on an Aboriginal and Torres Strait Islander Voice'. www.niaa.gov.au/indigenous-affairs/referendum-aboriginal-and-torres-strait-islander-voice, accessed 9 November 2023.

National Park Service (2015). 'Granite and Granodiorite FAQ'. www.nps.gov/goga/learn/education/granite-and-granodiorite-faq.htm, accessed 9 November 2023.

Neale, Margaret (2022). 'First Knowledges: An Introduction'. In Zena Cumpston, Michael-Shawn Fletcher and Lesley Head, eds, *Plants: Past, Present and Future*. Melbourne: Thames and Hudson, pp. 1–4.
Neimanis, Astrida (2016). *Bodies of Water: Posthuman Feminist Phenomenology*. London: Bloomsbury.
Neumark, Norie (2022). 'Squish, Squelch, shlshlshlurpppp'. In Sharon Jane Mee and Luke Robinson, eds, *Sound Affects: A User's Guide*. New York: Bloomsbury Academic, pp. 224–236.
Ngapulara Ngarngarnyi Wirra (Our Family Tree) (2022). Artwork by Old Ways, New.
Nicholson, Simon (2020). 'Solar Radiation Management'. *Wilson Centre*, 30 September. www.wilsoncenter.org/article/solar-radiation-management, accessed 12 December 2023.
Niman, Michael I. (2003). 'Kodak's Toxic Moments'. *Alternet*, May 28. www.alternet.org/2003/05/kodaks_toxic_moments, accessed 6 October 2023.
Nixon, Rob (2011). *Slow Violence and the Environmentalism of the Poor*. Cambridge, MA: Harvard University Press.
O'Hanlon, James (2023). *Silk & Venom: The Incredible Lives of Spiders*. Sydney: NewSouth Publishing.
Ocean Island Mine (2015). Performance Work by Latai Taumoepeau.
Odle, Melissa Patrice (2010). 'Bird Song in the Music of Olivier Messiaen, With an Emphasis on *Le Merle Noir*'. Master of Music in Performance Thesis, California State University, Northridge.
Oh! My Gulay! (2010). Performance Work, Cope/with/Land Theatre Company.
Old Ways, New (2023). Work Description. oldwaysnew.com/research-dev#spatial-awareness, accessed 14 December 2023.
Oli, Angus Nnamdi, Peter Anyigor Edeh, Reham M Al-Mosawi et al. (2020). 'Evaluation of the Phytoconstituents of *Auricularia auricula-judae* Mushroom and Antimicrobial Activity of Its Protein Extract'. *European Journal of Integrative Medicine*. 38: 101176.
Oliveri, Vicki Antonia, Glenn Porter, Chris Davies and Pamela James (2022). 'The Juukan Gorge Destruction: A Case Study in Stakeholder-Driven and Shared Values Approach to Cultural Heritage Protection'. *Journal of Cultural Heritage Management and Sustainable Development*, 1–15.
Pascoe, Bruce (2014). *Dark Emu, Black Seeds: Agriculture or Accident?* Broome, WA: Magabala Books.
Passengers (2016). Film, dir. Morten Tyldum. Columbia Pictures/LStar Capital/Village Roadshow Pictures/Wanda Pictures/Original Film/Company Films/Start Motion Pictures, United States.
Paterson, Eddie (2010). 'Paperless and Penless: Headphone Performance, Audioscripting and New Approaches to Writing for Performance'. In Catherine Cole, Marcelle Freiman and Donna Lee Brien, eds, *Strange Bedfellows: Refereed Conference Papers of the 15th Annual Australasian Association of Writing Programs Conference*, 1–9. www.aawp.org.au/publications/the-strange-bedfellows-or-perfect-partners-papers
Paterson, Eddie (2015). *The Contemporary American Monologue: Performance and Politics*. London and New York: Bloomsbury.
Paterson, Eddie (2019). 'Compost and Air-Conditioning: Beyond Biospherical Performance and towards the Shimmer'. *Performance Research*. 24(8): 31–36.
Paterson, Eddie and Kevin Brown, eds (2018) 'Performance Climates'. Themed issue. *Global Performance Studies* 1(2).
Phillips, Perdita (2001). *Termitaria*. Artwork.
Phillips, Perdita (2015). *Termite Embassy*. Artwork.
Philpott, Carolyn, Elizabeth Leane and Matt Delbridge, eds (2020). *Performing Ice*. Cham, Switzerland: Palgrave Macmillan, pp. 171–194.

Plumwood, Val (1993). *Feminism and the Mastery of Nature*. London: Routledge.
Plumwood, Val (1995). 'Human Vulnerability and the Experience of Being Prey'. *Quadrant*. 29(3): 29–34.
Plumwood, Val (2002). *Environmental Culture: The Ecological Crisis of Reason*. London: Routledge.
Plumwood, Val (2007a). 'The Cemetery Wars: Cemeteries, Biodiversity and the Sacred'. *Local-Global: Identity, Security, Community*. 3: 54–71.
Plumwood, Val (2007b). Review of Deborah Bird Rose's 'Reports from a Wild Country: Ethics for Decolonisation'. *Australian Humanities Review*. 42. www.australianhumanitiesreview.org/2007/08/01/a-review-of-deborah-bird-roses-reports-from-a-wild-country-ethics-for-decolonisation, accessed 22 November 2023.
Plumwood, Val (2012). *The Eye of the Crocodile*, ed. Lorraine Shannon. Canberra: ANU Press.
Ponyo (2008). Film, dir. Hayao Miyazaki. Studio Ghibli, Japan.
Pownall, Augusta (2019). 'Designers of Björk's Cornucopia Show "Ignored the Concert Rule Book"'. *De Zeen*, 13 December. www.dezeen.com/2019/12/13/bjork-cornucopia-chiara-stephenson-set-design/, accessed 9 November 2023.
Puig de la Bellacasa, Maria (2015). 'Making Time for Soil: Technoscientific Futurity and the Pace of Care'. *Social Studies of Science*. 45(5): 691–716.
Rancière, F., J. Botton, R. Slama et al. (2019). 'Exposure to Bisphenol A and Bisphenol S and Incident Type 2 Diabetes: A Case-Cohort Study in the French Cohort D.E.S.I.R.'. *Environmental Health Perspectives*. 127(10): 107013.
Readfearn, Graham (2021). 'World Heritage Committee Agrees Not to Place Great Barrier Reef on 'in Danger' List'. *The Guardian*, 24 July. www.theguardian.com/environment/2021/jul/23/world-heritage-committee-agrees-not-to-place-great-barrier-reef-on-in-danger-list, accessed 4 October 2023.
Reid, Robert (2019). 'The Detritus of Now: Escaped Alone'. Review for *Witness Performance*, 6 June. www.witnessperformance.com/the-detritus-of-now-escaped-alone, accessed 21 November 2019.
Remaud, Olivier (2022). *Thinking Like an Iceberg*, trans. Stephen Muecke. Cambridge: Polity Press.
Resources Victoria (2023). 'Victoria's Geology'. www.resources.vic.gov.au/geology-exploration/victorias-geology, accessed 9 November 2023.
Rinaldi, AC, O Comandini and TW Kuyper (2008). 'Ectomycorrhizal Fungal Diversity: Separating the Wheat from the Chaff'. *Fungal Diversity*. 33: 1–45.
Robot Unicorn Attack (2010). Computer Game, Spiritonin Media Games.
Rose, Deborah Bird (2017). 'Shimmer: When All You Love Is Being Trashed'. In Anna Tsing, Heather Swanson, Elaine Gan and Nils Bubandt, eds, *Arts of Living on a Damaged Planet: Ghosts and Monsters of the Anthropocene*. Minneapolis: University of Minnesota Press, pp. 51–63.
Rose, Deborah Bird (2017). 'Shimmer: When All You Love Is Being Trashed'. In Anna Tsing, Heather Swanson, Elaine Gan and Nils Bubandt, eds, *Arts of Living on a Damaged Planet: Ghosts and Monsters of the Anthropocene*. Minneapolis: University of Minnesota Press, pp. G51–G63.
Ross, Annabel (2016). 'Deaf Can Dance: The Delta Project Combines Hearing and Non-Hearing Dancers in Under My Skin'. Sydney Morning Herald, 29 April. www.smh.com.au/entertainment/dance/deaf-can-dance-the-delta-project-combines-hearing-and-nonhearing-dancers-in-under-my-skin-20160429-goinls.html, accessed 1 November 2023.
Roth, George B (1936). 'Dr. John Gorrie: Inventor of Artificial Ice and Mechanical Refrigeration'. *The Scientific Monthly*. 42(5): 464–469.
Sacrifice: Contemplating Time and Trust (2022). Performance Work. Science Gallery, University of Melbourne.

Saraceno, Tomás (2009). *Galaxies Forming Along Filaments Like Droplets on a Spider's Web*. Installation Work.

Saraceno, Tomás (2010). *14 Billions*. Installation Work.

Saraceno, Tomás (2022). *Webs of At-tent(s)ion*. Installation Work.

Schechner, Richard (2002). *Performance Studies: An Introduction*. London: Routledge.

Schneider, Rebecca (2016). 'Extending a Hand'. Keynote Address to Performance Climate, the 22nd Performance Studies International Conference. University of Melbourne.

Schneider, Rebecca and Paul Rae (2018). 'Extending a Hand/Leading an Ear'. *Performance Research*. 23(3): 13–24.

Serres, Michel (2016). *The Five Senses: A Philosophy of Mingled Bodies*. London: Bloomsbury.

Sheldrake, Merlin (2020). *Entangled Life: How Fungi Make Our Worlds, Change Our Minds and Shape Our Futures*. New York: Random House.

Shildrick, Margrit (2017). 'Border Crossings: The Technologies of Disability and Desire'. In Anne Waldschmidt, Hanjo Berressem and Moritz Ingwersen, eds, *Culture – Theory – Disability: Encounters between Disability Studies and Cultural Studies*. Bielefeld, Germany: Transcript, pp. 137–151.

Shlomowitz, Ralph and Doug Munro (1990). *The Ocean Island (Banaba) and Nauru Labour Trade: 1900–1940*. Bedford Park: Flinders University of Australia.

Silent Running (1972). Film, dir. Douglas Trumbull. Trumbull/Gruskoff Productions, United States.

Sloterdijk, Peter (2011). *Bubbles: Spheres Volume I: Microspherology*, trans. Wieland Hoban. South Pasadena CA: Semiotext(e).

Sloterdijk, Peter (2014). *Globes: Spheres Volume II: Macrospherology*, trans. Wieland Hoban. South Pasadena CA: Semiotext(e).

Sloterdijk, Peter (2016). *Foams: Spheres Volume III: Plural Spherology*, trans. Wieland Hoban. South Pasadena CA: Semiotext(e).

Spedalieri, Francesca (2014). 'Quietly Posthuman: Oriza Hirata's Robot-Theatre'. *Performance Research*. 19(2): 138–140.

Splash (1984). Film, dir. Ron Howard. Touchstone Films, USA.

Stabinsky, Doreen, Dipti Bhatnagar and Sara Shaw (2021). *Chasing Carbon Unicorns: The Deception of Carbon Markets and "Net Zero"*. Amsterdam: Friends of the Earth International.

Stelarc (2005). 'From Zombie to Cyborg Bodies – Extra Ear, Exoskeleton and Avatars'. *NeMe*, www.neme.org/texts/from-zombie-to-cyborg, accessed 22 November 2023.

Stelarc (2019). 'Extra Ear: Alternate Anatomical Architectures'. stelarc.org/_.php, accessed 11 December 2023.

Stelarc (2023a). 'Ear on Arm'. http://stelarc.org/_activity-20242.php, accessed 11 December 2023.

Stelarc (2023b). 'Extra Ear'. stelarc.org/_activity-20229.php, accessed 11 December 2023.

Stengers, Isabelle (2010). *Cosmopolitics*, vol. 1, trans. Robert Bononno. Minneapolis: University of Minnesota Press.

Stengers, Isabelle (2018). *Another Science Is Possible: A Manifesto for Slow Science*, trans. Steven Muecke. Cambridge: Polity Press.

Stevens, Lara (2019a). 'Anthroposcenic Performance and the Need For "Deep Dramaturgy"'. *Performance Research*. 24(8): 89–97.

Stevens, Lara (2019b). *Not Now, Not Ever*. Performance Work.

Stevens, Lara (2022). 'Sink or Swim: Performing the Iniquities of the Climate Crisis'. *Critical Stages/Scènes Critiques*. (26). www.critical-stages.org/26/sink-or-swim-performing-the-iniquities-of-the-climate-crisis.

Stevens, Lara and Denise Varney (2022). 'The Climate Siren: Hanna Cormick's *The Mermaid*'. *TDR: The Drama Review*. 66(3): 107–118.

Strandvad, Sara Malou, Tracy C Davis and Megan Dunn (2021a). 'Mermaids as Market Creators: Cultural Entrepreneurship in an Emerging Practice'. *International Journal of Cultural Studies*. 25(1): 68–85.

Strandvad, Sara Malou, Tracy C Davis and Megan Dunn (2021b). 'Skills and Strategies of Activist Mermaids: From Pretty to Powerful Pictures'. *Text and Performance Quarterly*. 41(3–4): 262–282.

Stukenborg, JB, RT Mitchell and O Söder (2021). 'Endocrine Disruptors and the Male Reproductive System'. *Best Practice & Research: Clinical Endocrinology & Metabolism*. 35(5): 101567.

Sundar, Pratap, Chandan Chowdhury and Sagar Kamarthi (2021). 'Evaluation of Human Ear Anatomy and Functionality by Axiomatic Design'. *Biomimetics*. 6(2): 31.

Sunshine (2007). Film, dir. Danny Boyle. Moving Picture Company/DNA Films/UK Film Council/Ingenious Film Partners, United Kingdom.

Suz, LM, S Kallow, K Reed et al. (2017). 'Pine Mycorrhizal Communities in Pure and Mixed Pine-Oak Forests: Abiotic Environment Trumps Neighboring Oak Host Effects'. *Forest Ecology and Management*. 406: 370–380.

Tait, Peta (2019). 'Striving, Falling, Performing: Phenomenologies of Mood and Apocalypse'. In Peter Eckersall and Helena Grehan, eds, *The Routledge Companion to Theatre and Politics*. London: Routledge, pp. 284–287.

TallBear, Kim (2014). 'Standing with and Speaking as Faith: A Feminist-Indigenous Approach to Inquiry'. *Journal of Research Practice*. 10(2): 1–7.

Teaiwa, Katerina Martina (2015). *Consuming Ocean Island: Stories of People and Phosphate from Banaba*. Bloomington and Indianapolis: Indiana University Press.

Teaiwa, Katerina Martina (2021). 'Katerina Teaiwa'. *VCA Art Forum*. 8 October. https://www.youtube.com/watch?v=fFJQC1W-qDI, accessed 12 December 2023.

Teaiwa, Teresia K (1994). 'Bikinis and Other s/pacific n/oceans'. *The Contemporary Pacific*. 6(1): 87–109.

Terrain (2012). Performance Work, Bangarra Dance Theatre.

The Australian Dream (2019). Documentary Film, dir. Daniel Gordon. Good Thing Productions/Lorton Entertainment/Passion Pictures, Australia.

The Constant State of Desire (1986). Performance Work by Karen Finley.

The Final Quarter (2019). Documentary Film, dir. Ian Darling. Shark Island Productions, Australia.

The Little Mermaid (1989). Film, dir. John Musker and Ron Clements. Walt Disney Pictures, USA.

The Martian (2015). Film, dir. Ridley Scott. Scott Free Productions/Kinberg Genre/TSG Entertainment, United Kingdom.

Theatrum Botanicum (2022). Performance Work by Gregory Lorenzutti.

Tsing, Anna Lowenhaupt (2015). *The Mushroom at the End of the World: On the Possibility of Life in Capitalist Ruins*. Princeton, NJ: Princeton University Press.

Tubiana, Laurence, Bruno Latour and Frédérique Aït Touati (2015). Paris *Climate – Make It Work*. Performance Event.

Uluru Statement (2023). 'Uluru Statement from the Heart'. www.ulurustatement.org/the-statement/view-the-statement/, accessed 9 November 2023.

Under My Skin (2016). Performance Work, The Delta Project.

Unicorn, Gratitude, Mystery (2016). Performance Work by Karen Finley. University Press.

van Baarle, Kristof (2019). 'Void'. In Peter Eckersall and Helena Grehan, eds, *The Routledge Companion to Theatre and Politics*. London: Routledge, pp. 228–231.

van Baarle, Kristof (2020). 'End-Time Attitudes: Performing the Last Part'. In Kristof van Baarle and Peter Eckersall, eds, *Machine Made Silence: The Art of Kris Verdonck*. Aberystwyth, UK: Performance Research Books, pp. 152–162.

van Vegten, Megan (2022). 'Standing Stones and Swarm Robotics... Artist Dr Robert Walton Explores the Mysteries of Old and New Technology'. *Ingenium*, Faculty of Engineering and Information Technology, University of Melbourne, 30 September https://eng.unimelb.edu.au/ingenium/standing-stones-and-swarm-robotics-artist-dr-robert-walton-explores-the-mysteries-of-old-and-new-technology, accessed 9 November 2023.

Varney, Denise and Lara Stevens (2023). 'Australian Coal Theatrics'. *TDR: The Drama Review*. 67(1): 57–61.

Varney, Denise, Peta Tait and Lara Stevens (2021–2024). *Towards An Australian Ecological Theatre*. Australia Research Council Discovery Project, University of Melbourne.

Verdonck, Kris (2016). *IN VOID*. Performance Work, A Two Dogs Company.

Verdonck, Kris (2017). *CONVERSATIONS (at the end of the world)*. Performance Work, A Two Dogs Company.

Verdonck, Kris (2019). *SOMETHING (out of nothing)*. Performance Work, A Two Dogs Company.

Verdonck, Kris (2023). *PREY*. Performance Work, A Two Dogs Company.

Verdonck, Kris and Kristof van Baarle (2022). 'Between Machine and Prey'. Artist talk delivered at CLIMARTE Gallery, Richmond, Victoria, 26 May.

Viney, Steven and Edwina Seselja (2022). 'Postcards from the Frontlines of Climate Change'. *ABC News*, 11 May. www.abc.net.au/news/2022-05-11/climate-change-effects-facts-causes-pacific-postcards/100999144, accessed 12 October 2023.

Waddoups, Ryan (2022). 'Björk's New Album Is an Ode to Mushrooms'. *Surface Magazine*, 3 October. www.surfacemag.com/articles/bjork-fossora-album-cover/, accessed 9 November 2023.

Wadrill, Violet, Topsy Dodd Ngarnjal, Leah Leaman et al. (2023). *Tamarra: A Story of Termites on Gurindji Country*. Melbourne: Hardie Grant.

Wahlquist, Calla (2020). 'Australia's Summer Bushfire Smoke Killed 445 and Put Thousands in Hospital, Inquiry Hears'. *The Guardian*, 25 May. www.theguardian.com/australia-news/2020/may/26/australias-summer-bushfire-smoke-killed-445-and-put-thousands-in-hospital-inquiry-hears, accessed 14 September 2023.

Waiting (2017). Art installation, Out-of-Sync.

Watson, Irene (2014). 'Re-Centring First Nations Knowledge and Places in a Terra Nullius Space'. *AlterNative: An International Journal of Indigenous Peoples*. 10(5): 508–520.

Ważny, R (2014). 'Ectomycorrhizal Communities Associated with Silver Fir Seedlings (*Abies alba* Mill.) Differ Largely in Mature Silver Fir Stands and in Scots Pine Forecrops'. *Annals of Forest Science*. 71: 801–810.

Welton, Martin (2012). *Feeling Theatre*. Basingstoke: Palgrave Macmillan.

Williams, David (2015). 'Epilogue: "The Ruins of Time (I've Forgotten This Before)"'. In Carl Lavery and Clare Finburgh, eds, *Rethinking the Theatre of the Absurd: Ecology, the Environment and the Greening of the Modern Stage*. London: Bloomsbury, pp. 241–250.

Wilson, Alex (2008). '"N'tacinowin Inna Nah": Our Coming in Stories'. *Canadian Woman Studies*. 26(3): 193–199.

Wilson, Richard (2015). 'A Globe of Sinful Continents: Shakespeare Thinks the World'. In Jernej Habjan and Fabienne Imlinger, eds, *Globalizing Literary Genres: Literature, History, Modernity*. New York: Routledge, pp. 27–49.

Wilson, Shawn (2008). *Research Is Ceremony: Indigenous Research Methods*. Black Point, Nova Scotia: Fernwood Publishing.

Wright, Alexis (2023). 'The Black Summer Bushfires Killed 3 Billion Animals. They Are Our Relatives; They Deserve to be Mourned'. *The Guardian*, 1 April. www.theguardian.com/australia-news/2023/apr/01/alexis-wright-black-summer-bushfires-vigil-sydney-speech-3-billion-animals-killed-are-our-relatives-they-deserve-to-be-mourned, accessed 12 June 2023.

Wright, Alexis (2023). *Praiseworthy*. Sydney: Giramondo.
Wudjang: Not the Past (2022). Performance Work, Bangarra Dance Theatre.
WWF-Australia (2020). 'Australia's 2019–2020 Bushfires: The Wildlife Toll'. wwf.org.au/blogs/australias-2019-2020-bushfires-the-wildlife-toll/, accessed 27 January 2023.
Yalcinkaya, Gunseli (2018). 'Olafur Eliasson Installs Giant Blocks of Glacial Ice Across London'. *Dezeen*, 12 December. www.dezeen.com/2018/12/12/ice-watch-olafur-eliasson-installation, accessed 12 October 2023.
Young, Davis (2003). *Mind Over Magma: The Story of Igneous Petrology*. Princeton: Princeton University Press.
Zhang, Rui Ying, Dan Dan Hu and Jin Gang Gu, et al. (2012). 'Development of SSR Markers for Typing Cultivars in the Mushroom *Auricularia auricula-judae*'. *Mycological Progress*. 11: 587–592.
Zuboff, Shoshana (2019). *The Age of Surveillance Capitalism*. London: Profile Books.
Zylinska, Joanna, ed (2002). *The Cyborg Experiments: The Extensions of the Body in the Media Age*. London: Continuum.

Index

FSC
www.fsc.org
MIX
Papier | Fördert gute Waldnutzung
FSC® C083411

Kolibri 360 GmbH
Ferdinand-Jühlke-Straße 7
99095 Erfurt, Deutschland
produktsicherheit@kolibri360.de